Introduction to Show Control

Connecting Entertainment Control Systems for Live Shows

JOHN HUNTINGTON

ZIRCON DESIGNS PRESS

Copyright © 2023 by John Huntington, All Rights Reserved

No part of this book may be reprinted or reproduced or utilized in any form or by any electronic, mechanical, or other means, now known or hereafter invented, including photocopying and recording, or in any information storage or retrieval system, without permission in writing from the publisher.

Although every precaution has been taken to verify the accuracy of the information contained herein, the author and publisher assume no responsibility for any errors or omissions. No liability is assumed for damages that may result from the use of information contained within.

Publisher:
Zircon Designs Press
NYC, USA
www.zircondesigns.com

Print ISBN-13: 978-1-7357638-4-2

EBook ISBN-13: 978-1-7357638-5-9

BISAC: Performing Arts / Theater / Stagecraft

Version 1.0, May 15, 2023

TABLE OF CONTENTS

PREFACE
Roots Of This Book. xiii
For Whom Is This Book Written?. xv
Conventions. xv
Disclaimers. xv
Thanks to My Production Team xv
Website and Contact Information xv

CHAPTER 1
INTRODUCTION
What Is Show Control?. 1
    Networks 1
    Programming 3
    Benefits of Show Control. 3
    The Process of Show Control. 4
    How are Show Control Systems Used? 4
Cues 4
Show Types 5
    Linear Shows 5
    Nonlinear Shows 5
Triggers and Synchronization 6
    Event Based 6
    Time Based 7
    Hybrid Systems 8

Cueing Concepts 9
    Open and Closed Loop 9
    System Flexibility. 10
    Absolute and Relative. 10
Show Control System Architectures 13
    Primary-Secondary, Leader-Follower. 13
    Peer-to-Peer 14
    Hybrid 14
    Centralized or Distributed Systems 14
    Other Structures. 15
Entertainment Control Disciplines and Devices 16
    Lighting 16
    Audio 16
    Video 17
    Lasers 17
    Stage Machinery. 17
    Animatronics 17
    Fog, Smoke, Fire, and Water. 18
    Pyrotechnics 18
    Building Systems 18
    Other Show Controllers. 18
Moving On 19

CHAPTER 2

CONNECTING DEVICES AND SYSTEMS
Data Communications. 21
    Legacy Point to Point Standards 21
    Networking. 22

- Other Connection Methods 22
- Control Protocol Approaches 22
- Custom Network-Based Protocols 24
  - ASCII 24
  - Network Details 25
- Open Sound Control (OSC) 25
  - Clients and Servers 25
  - Addresses 26
  - Examples 26
  - OSC in Show Control 27
- Musical Instrument Digital Interface (MIDI) 27
  - MIDI Messages 28
  - MIDI Hardware 28
  - MIDI Applications in Show Control 29
- MIDI Show Control (MSC) 29
  - Targeting Devices 29
  - Cue Data 30
  - Some Examples 30
  - Limitations of MIDI Show Control 31
- Time Code 32
  - Show Control Time Code Applications 33
  - Frame Rates 33
  - SMPTE Time Code 35
  - MIDI Time Code (MTC) 36
  - Practical Time Code For Live Shows 36
  - Time Code Over a Network 38
- Other Ways to Synchronize Systems 38
  - Running Wild 38

Proprietary Time Codes................................39
Click Track..39
Digital MultipleX (DMX512).............................39
DMX Over a Network..................................40
Remote Device Management (RDM)..........................40
Contact Closures for Input or Output........................41
Contact Closure Inputs...............................41
Direct Outputs....................................44
A Note On Electrical Isolation..........................45
Miscellaneous Protocols................................46
MQTT..46
Positional Tracking and Interchange Protocols..................46
Simple Network Management Protocol (SNMP)................46
Network Time Synchronization Protocols....................46
Moving On...47

CHAPTER 3

A SHOW CONTROL DESIGN PROCESS

System Design Principles................................49
1. Ensure Safety...................................49
2. The Show Must Go On.............................53
3. Simpler Is Always Better............................56
4. Strive for Elegance................................56
5. Complexity Is Inevitable, Convolution Is Not................56
6. Make It Scalable, Leave Room for Unanticipated Changes.......56
7. Ensure Security..................................57
Show Control System Design Process........................58
What are the safety considerations?.......................58

  What type of show is it? 59
  What kind of triggers and synchronization should be used? 59
  What devices or systems are to be connected/controlled? 59
  What system architecture works for this application? 59
  What is the control information source? 59
  What type of user interface is required? 60
  What kind of show control approach is needed? 60
 A Note On Budget and Time 62
 Do You Really Need Show Control? 62
 Moving On 63

CHAPTER 4

EXAMPLE SHOW CONTROL SYSTEMS

 Entering The Arena 65
  The Mission 65
  What are the safety considerations? 65
  What type of show is it? 66
  What kind of triggers and synchronization should be used? 66
  What devices or systems are to be connected/controlled? 66
  What system architecture works for this application? 66
  What is the control information source? 66
  What type of user interface is required? 66
  What kind of show control approach is needed? 67
 A Theatrical Thunderstorm 69
  The Mission 69
  What are the safety considerations? 69
  What type of show is it? 70
  What kind of triggers and synchronization should be used? 70

What devices or systems are to be connected/controlled?	70
What system architecture works for this application?	72
What is the control information source?	72
What type of user interface is required?	72
What kind of show control approach is needed?	72

Ten-Pin Alley... 76
    The Mission.. 77
    What are the safety considerations?........................... 77
    What type of show is it?..................................... 78
    What kind of triggers and synchronization should be used?...... 78
    What devices or systems are to be connected/controlled?........ 78
    What system architecture works for this application?........... 80
    What is the control information source?....................... 81
    What type of user interface is required?...................... 81
    What kind of show control approach is needed?................. 81

Comfortably Rich 84
    The Mission.. 85
    What are the safety considerations?........................... 85
    What type of show is it?..................................... 85
    What kind of triggers and synchronization should be used?...... 85
    What devices or systems are to be connected/controlled?........ 85
    What system architecture works for this application?........... 87
    What is the control information source?....................... 87
    What type of user interface is required?...................... 87
    What kind of show control approach is needed?................. 87

Popsicle Wolf.. 90
    The Mission.. 90
    What are the safety considerations?........................... 90

    What type of show is it?......................................90

    What kind of triggers and synchronization should be used?........90

    What devices or systems are to be connected/controlled?..........91

    What system architecture works for this application?..............92

    What is the control information source?.......................92

    What type of user interface is required?........................92

    What kind of show control approach is needed?..................93

It's an Itchy World after All...................................94

    The Mission..95

    What are the safety considerations?...........................95

    What type of show is it?......................................95

    What kind of triggers and synchronization should be used?........95

    What devices or systems are to be connected/controlled?..........95

    What system architecture works for this application?..............98

    What is the control information source?.......................98

    What type of user interface is required?........................98

    What kind of show control approach is needed?..................98

CONCLUSION

    Contact Info and Blog.......................................107

    Index..109

PREFACE

ROOTS OF THIS BOOK

In 1990, I was a Technical Editor at *Theatre Crafts* and *Lighting Dimensions* magazines here in New York, and fresh out of grad school, where I had written my thesis on entertainment control systems and show control. It was clear to me, all those decades ago, that digital control communication technologies were increasingly forming the backbone of entertainment systems, and one day, looking for technical references while researching an article, I went to Times Square to visit the Drama Book Shop (the Amazon.com of the day for books on entertainment in that pre-Google era). On the shelves, I found a wide variety of show technology titles that explained scenic construction, lighting design principles, paint techniques, and even sound in those early days of digital audio. But I wasn't able to find a single book that explained the then-widely used DMX512 digital lighting control standard, and I was even more shocked to find that the majority of the lighting technology books didn't even *mention* it. I also didn't find anything about the details of show control technologies like SMPTE Time Code or MIDI, both of which had been in the market for years by that point. That bookstore visit started a long process that led to the 1994 release of my book *Control Systems for Live Entertainment*, the earliest ancestor of this *Introduction to Show Control*, released almost 30 years later.

Focal Press was my publisher for that first edition of *Control Systems for Live Entertainment*, and released my updates in 2000 and 2007. But by 2011—after thousands of copies sold—they decided they were no longer interested in the book. Additionally, around that time, the live entertainment industry hit a big technological maturity point[1], and also seemed to be in a similar technical information situation as we had been back in 1990, but this time regarding networks and their ever-increasing use on shows. Searching Amazon and Google for things like "show networking" in those days, I found mostly references back to my own book, along with others that didn't really give a complete picture of the ways networks were used on shows. I found this to be a bit of surprising déjà vu, since networks increasingly were at the core of entertainment and show control systems. I felt a new edition of the book

---

1 See my blog for links to two articles I wrote about this maturity in 2019. https://www.controlgeek.net/blog/2020/8/20/development-and-evolution-of-show-technology-articles-and-timeline

with an expanded networking focus was warranted, so I completely reorganized the content, added a lot on networking (the first 1994 edition had about fifteen pages on networking; by this edition I was up to three full chapters), and self published the first edition of the re-titled *Show Networks and Control Systems* in 2012.

As the market continued to mature, I released an update in 2017[2], where I actually started removing now-irrelevant information from the book, as Ethernet became the unified digital data highway for entertainment technology. It became clear that most entertainment technicians needed to know something about networking, but didn't necessarily need all the information in the 475 page, 175,000 word behemoth that *Show Networks and Control Systems* had become. Additionally, many of the low-level details (connector pin outs, binary message breakdowns, etc.) covered in hundreds of pages in my original books became packaged into software and products (with details easily found online when needed), and made ever easier to use by the brilliant engineers working for our manufacturers.

So, around 2019, I decided to retire much of the low-level and historical detail from the "big" book, and take the still-relevant material and break it into two smaller books: one on show networking for the broader entertainment technology market, and another on the niche of show control. With some time during the COVID lock down, I completed the first phase of that work, extracting out, updating, reorganizing, and expanding the networking information, and released the first edition of the smaller, more focused, self-published *Introduction to Show Networking* in 2020. And now, in 2023, I'm completing the process by releasing *Introduction to Show Control*. With the publication of this book, I'm ending my almost 30 year attempt to comprehensively—in a single volume—survey the entertainment control, show control, and networking market. And going forward, I'm leaving much of the low-level detail to other resources, and have moved historical asides and other original, interesting, but no longer germane topics from the "big book" onto my blog[3].

---

2   I made one last small updated edition in 2021, after my book got marooned by my printer. I did a presentation on this at the Hackers On Planet Earth conference in 2022; link on my blog https://www.controlgeek.net/blog/2022/8/18/my-moving-beyond-amazon-self-publishing-purgatory-talk-from-a-new-hope-hackers-on-planet-earth-conference

3   https://www.controlgeek.net/blog/category/Entertainment+Technology

FOR WHOM IS THIS BOOK WRITTEN?

Show control by definition (page 1) is connecting more than one entertainment control system together. And as such, show control is an upper level topic, so I'm assuming here that readers have some background in entertainment control systems. However, while of course I don't expect readers to be experts in all of it, they should be familiar with the gear found on a show (lighting, sound, video, etc.), and basic show production processes and techniques.

CONVENTIONS

If a term is **bolded**, then it is a "key" term; I try to mark the first major usage of the term in the book. Additionally, there are a number of cross references throughout the book. In print, they should refer to a page number; in electronic form they should contain a link to take you to the related part of the book. I've generally only included forward cross references here when speaking about something we haven't covered yet; to look backward, there is a detailed Table of Contents (page v) and an Index (page 109).

DISCLAIMERS

And now for the *"It's not my fault!"* disclaimer: While I've made every effort to ensure that the information in this book is accurate, DO NOT implement anything in any product or system based solely on the information in this book. The goal here is understanding; if you want to go to the next level—*implementation*—you need to obtain information from the appropriate standards or other organizations. Additionally, safety is the responsibility of system designers and operators. I include some general safety principles in this book; these are based on the way I do things and not necessarily in compliance with any industry standard. It is *your* responsibility to ensure safety in any system with which you deal!

THANKS TO MY PRODUCTION TEAM

Literally hundreds of people helped me with this book and its predecessors over the years. But I want to extend a special thanks to Aaron Bollinger for creating all the excellent illustrations in the book over many years; Shelbye Reese, who designed the fantastic layout and the cover, and Michael Lawrence for copy editing.

WEBSITE AND CONTACT INFORMATION

Errata for this book, my blog, a contact form, and much more is available on my website: http://www.controlgeek.net

Chapter 1

INTRODUCTION

In show business, storytelling has been our job for thousands of years. Entertainment technology provides a powerful set of storytelling tools that have extended far beyond their roots in traditional performing arts like theater, and out into a wide array of venues and types of shows, including concerts, circuses, theme parks, corporate meetings, wrestling shows, special events, cruise ship shows, themed retail, "immersive" art installations, fountain spectaculars, museum exhibits, mega churches, and so on. And those kinds of shows—presented to an audience in the same room at the same time—are what we will be focusing on here.

WHAT IS SHOW CONTROL?

Show control simply means connecting two or more **entertainment control systems** together. Entertainment control systems are used in many elements or disciplines in the show environment; for example, the control signals used between a lighting control console and a moving light; the network used to synchronize and control a number of video servers; the digital data sent between a pyro controller and a flash pot firing system; or anything else related to the field. A computer controlling fog machines that regulates the amount of fog in a harbor scene doesn't amount to show control; a system that *links* the control of the fog machine with an audio playback system generating maritime sound effects does. A concert lighting system may be made up of a large network and a sophisticated control system, but that's entertainment control. If you synchronize the lighting console with a video clip, that's show control. The key is that you don't have a *show control system* unless control for *more than one* production element is linked *together*.

Sophisticated show control systems often have two key aspects that we aren't covering here in this book: networking and programming. Let's discuss why they aren't included here before moving on.

Networks

Networking is a critical part of show control, entertainment control, and show media distribution, and thereby plays an important role on most live shows. Because networking has applications well beyond show control, I have written extensively about it in a separate book: *Introduction to Show Networking*[1]. I have designed this *Introduction to Show Control* book to be a companion volume to the networking book; since networking is so important for show control, you likely want to read *Introduction to Show Networking* before moving on to the more advanced sections here.

But we should get a couple basic definitions on the table, so here's an excerpt[2] from *Introduction to Show Networking*:

> A **network** is two or more devices using a common physical infrastructure to allow each connected computer to communicate with all the others. Any device connected by the network is called a **node**; if the node has data to communicate, it may be referred to as a **host**. **Ethernet**, the most widely-used network standard, offers incredible flexibility and power at a low cost, and—built correctly—Ethernet networks are robust, reliable and perform many mission-critical functions on our shows.
>
> Networks are found on shows of all sizes, primarily serving two roles. First, networks transport control data to operate show equipment (lights, sound playback systems, video servers, automated scenery controllers, rigging controllers, pyro and special effect devices, etc.); this is a critical function they have been serving since the 1990s. Additionally, once network capacity increased, networks became widely used to transport, or "stream[3]", digital audio and video media. The real power of Ethernet is that—built correctly—the same network could work well in either role, carrying just about any kind of **digital** data for a show.

In order to communicate with a device on a network, everything needs an address, and the **IP Address** is one other key concept from show networking. Here's another excerpt[4] from the *Introduction to Show Networking* book.

> To be delivered properly to its destination, a postal letter needs an address unique to that destination. Similarly, each packet of data on a network

---

1   Details on how to get the networking companion volume here: https://www.control-geek.net/bookinfo

2   Page 1.of the first edition of the networking book.

3   The term "streaming" is generally more associated with online Internet transport of media; we often operate on closed networks not connected to the larger Internet..

4 Page 36 of the first edition of the networking book.

needs a destination address, and handling this information is one of the key functions of the **Internet Protocol** (IP). IP version 4 (IPv4) is the backbone of the vast majority of show networks, and it provides a universal addressing scheme that can work within or between a wide variety of networks (hence the "inter" name), providing unique IDs for the connected hosts.

Finally one more networking term to define that we will need later: bandwidth.[5]

A digital data link carries a binary stream of 1s and 0s. The rate of transmission is known as the **bit rate**, which is measured in **bits per second** (bit/s, or BPS). Bit rate measurements use the International System of Units (SI) prefixes, so it's very common to see something like Mbit/s, a megabit (1 million) per second, or Gbit/s, a gigabit (1 billion) per second. Whatever the transmission medium, there is always some limit as to how much data a single communications connection can handle; this capacity is known as the channel's **bandwidth**. A "high bandwidth" connection can carry more than a "low bandwidth" link.

Programming

Programming and coding are important aspects of show control that I reference here but do not cover in detail, because the languages and syntax used in show control programming are often very closely tied to the systems being programmed. So, while the ideas of algorithms are common and we can discuss those in a broad sense here, I'm leaving the coding details to numerous other, specific resources.

Benefits of Show Control

There are many reasons to connect entertainment control systems together, but the key reasons for me are:

- To increase cueing and synchronization precision beyond human capability.
- To maximize the productivity of available labor.
- To allow the performers to interact with the technical elements (if desired).

We'll explore all this in further detail as we move forward.

---

5   Page 6 of the first edition of the networking book.

The Process of Show Control

I find designing and implementing show control to be a fascinating endeavor, and it suits me because I find myself to be a "general specialist": I know something about a lot of different things but I'm not really an expert in any one area. Show control is about connecting with different departments and areas, so while expertise in one area is valuable (and the more the better—I'm primarily a sound person), being able to know a little about other systems on a show is helpful. Employing a "systems thinking" mindset is also helpful; and of course networking knowledge is critical. Strong communications skills and the ability to be diplomatic are also important, since we are often crossing department lines (basically, be nice!). But probably the most useful thing to have in this field is an innate curiosity. When I hang out with fellow show control geeks, our conversations run a wide gamut; I think that curiosity about a lot of different things and all kinds of systems throughout the world is what draws us to this weird niche.

How are Show Control Systems Used?

Show control applications abound throughout live show production and a variety of kinds of shows and installations, ranging from a small show to a huge stadium or theme park spectacular. In some cases, show control can be implemented with a single cable run between two controllers providing timing information; alternatively, show control could be made up of a sophisticated network-based system that controls and connects an entire production or installation. In Chapter 4 on page 65 we will work through a design process for six different examples; you might keep these in mind as you move through the rest of this book. In "Entering The Arena" on page 65 we will go through an audio/video/lighting show for a sports team's entrance. Next in "A Theatrical Thunderstorm" on page 69 we will go through a lighting/sound/effects example from the theatrical world. Then on "Popsicle Wolf" on page 90 we will go through an "immersive" attraction example, and then move onto a concert lighting/video/sound application with "Comfortably Rich" on page 84. After that is a "themed retail" example in "Ten-Pin Alley" on page 76, covering animatronics and an audio/video/lighting show and then we close with an interactive stunt show theme park example in "It's an Itchy World after All" on page 94.

CUES

The **cue** is the basic building block of show production, and it's a concept we deal with so intuitively that we may not have ever really thought about a definition. Computers and networks, however, like things to be defined precisely, and some readers may not be familiar with the field, so let's begin with a general introduction.

One definition of cue from the American Heritage Dictionary is: "A signal, such as a word or action, used to prompt another event in a performance, such as an actor's speech or entrance, a change in lighting, or a sound effect.[6]" While this definition refers to the signal for an action itself, in our world backstage a cue could refer to a number of different things, depending on the production element to which one is referring and its current state and context. For example, a cue can be a self-contained event that once started runs until completion, like pyro cue 76, which triggers a flash effect (once started, the pyro cue can't be stopped). A cue can also modify something that is continuing: a lighting cue is generally a transition from one light "look" to another, but "light cue 18" can refer to both the *transition from* light cue 17 *and* the resulting new look 18, which will continue onstage after execution of the cue transition.

SHOW TYPES

In order to design a show control system, we have to first think a bit about the type of show for which the system will be used, because this often impacts design decisions. For our purposes, I'm going to categorize all shows into two basic types: linear and nonlinear.

Linear Shows

A **linear show** is one in which there is a single, fixed storyline that is normally performed in the same sequence in each show. This may take the form of a live theatrical-style show, where the performers act out the same script in each show; a cruise ship revue or a concert where the same songs are played in order each performance; or a fountain and laser show, where the sequence is prerecorded. For these kinds of shows, show and entertainment control systems can be designed so that cues are generally executed in the same order each performance.

Nonlinear Shows

In a **nonlinear show**, multiple, separate components of a show can all run independently, in a dynamic order, or even simultaneously. For example, a band that plays a concert but does not have a rigid, defined set list would be nonlinear, since the songs are not necessarily played in the same order each night. Another example would be a haunted house attraction, where the audience members themselves activate sensors, triggering various aspects of the show as they walk through. For

---

6 *The American Heritage® Dictionary of the English Language, Fifth Edition copyright ©2022 by HarperCollins Publishers.* https://www.ahdictionary.com/word/search.html?q=cue

these kinds of shows, show control systems have to be designed so that the cues can be executed in any order; this demands some structural flexibility.

TRIGGERS AND SYNCHRONIZATION

Cues and show elements must somehow be triggered or synchronized. Generally, triggering approaches fall into three categories: event based, time based, or hybrids of the two. Note that these concepts are related to the ideas of linear and nonlinear shows, but this is not a fixed relationship.

Event Based

Theatrical-style live performances are generally run in an event-based way, with the cues triggered based on the occurrence of events not necessarily fixed in time. For instance, when a performer screams, "Run for your life!," a cue might be triggered to create explosion lighting and sound effects. The event of the performer's utterance of the line is the trigger for the cues. Event-based systems allow performers to vary their timing or improvise, and these systems can (with enough intelligence in the system) flexibly accommodate mistakes.

Here are some examples of events that could be used to trigger cues:

- A performer speaks a line of text.
- A performer sits on an onstage sofa.
- The sun appears over the horizon[7].
- A specific measure and beat of music is performed.
- A drummer hits a drum.
- An operator presses a button on a control console.
- An audience member passes a sensor

Event-Based Cue Lists

Event-based systems are most typically represented in show computer systems by "Cue Lists," with the cues listed out vertically. Here, the "standing by" cue is indicated by the gray bar—this indicates which cue ("Wind") will be executed when the "Go" button is pressed:

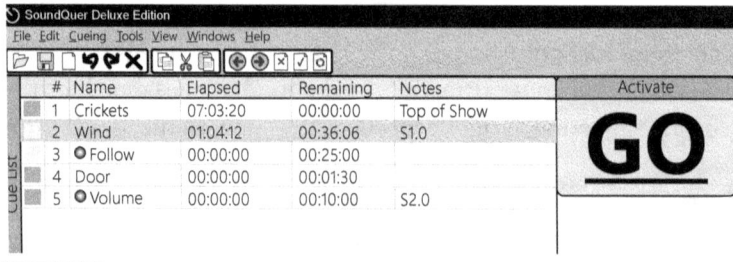

---

7 Of course we can predict the time of the sunrise, but here we are waiting to actually see it.

Time Based

A fireworks display set to music, a performance with prerecorded backing music tracks, an animatronic character performance, and a televised half-time spectacular are all examples of show types that typically run their show elements using time-based methods.

There are two general types of time-based triggering or synchronization. In one approach, all elements, including human performers, are synchronized by some connection to a clock; cues (events) are then fired at specific, preprogrammed times. Alternatively, two or more types of time-based linear media (audio recording, video clip, or similar) can be synchronized to each other (or to a separate clock) on an ongoing basis.

A time-based system is less forgiving to human performers. If, for example, the explosion effects described above were triggered at 4 minutes and 35 seconds into every performance, the performer would be out of luck if their performance varied much. Alternatively, if a performer forgets their lines or trips and falls, the show will go on without them, and it's up to the performer to catch up. With these systems, the performer becomes simply another follower of an automated conductor, synchronizing themselves to the control system, instead of the other way around. However, (to generalize) time-based show control systems have some advantages over event-based systems: They are very straightforward (although time consuming) to program, and then easy to run automatically, predictably, and reliably, eliminating many variables. Such a system may sound limiting, but shows of all kinds have successfully run this way for many years.

Typically, in time based systems, the clock comes from (or is synchronized to) some sort of preprepared media that is already constrained and fixed in time, such as a video segment or audio sound track. There are many ways to synchronize all the other show elements to this time clock; for example, some sort of **time code** (page 75) can be sent from one device to all the controlled devices. Alternatively, all the systems can be told to start at exactly the same time, and then each system runs **wild** (page 38), with no continuing re-synchronization.

Here are some examples of time-based show applications:

- The show starts at 3 p.m. Eastern Standard Time.
- A light cue is triggered at 10 seconds after 3 p.m.
- Two minutes, 22 seconds, and 12 frames into the video, a strobe is fired.
- Four minutes and 15 seconds into the show's audio sound track, a fireworks

shell is fired.
- Twenty-minute long audio and video clips are continuously synchronized together.

Time-Based Timelines and Cue Lists

Time-based systems are often represented in computer systems by "timelines," with the events or cue media displayed horizontally. The current point in time is traditionally indicated by a horizontally scrolling vertical bar or pointer:

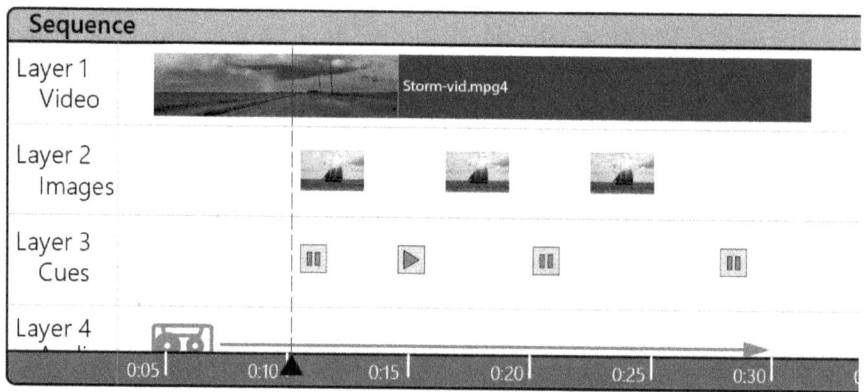

Alternatively a sequence of time-based cues can be represented in a list with cue trigger times:

Time	Cue	Action
00:00:00	1	Houselights Out
00:00:05	2	Music 1 Starts
00:00:09	3	Stage Lights Up
00:00:16	4	Video Start
00:01:00	5	Sound FX 1
00:01:10	6	Sound FX 2
00:02:13	7	Video Pause
00:02:15	8	Music 1 Cuts Out
00:03:00	9	Music 2 Starts
00:03:10	10	Lights Out

Hybrid Systems

Many show control systems are hybrids of event- and time-based systems. Systems that are primarily event based can have built-in time-based sequences; conversely, a time-based system can stop and wait for an event trigger before continuing. In our example above, if the explosion effects were built into a five-second-long time-based sequence, the system would then be hybrid. Most of the show would be event based, and when the actor yells, "Run for your life!", a preprogrammed, time-based sequence is initiated, triggering the lighting and sound cues for the explosion in a precise, repeatable way. The actors synchronize their movements to the time-based sequence, which is relatively straightforward, since the sequence runs exactly the same way every night. At the conclusion of the time-based sequence (itself an event), the show returns to an event-based operating mode and cues continue based on the script, an actor's position on the stage, or other events.

An additional type of hybrid synchronization is **musical time**, where systems are synchronized to musical songs, measures, and beats. The duration of the song or even an individual beat can vary based on the performance, so it's not really time based; since it's made up of beats and measures it's not really purely event based. Cues are often synchronized in this way in opera, musicals, and similar types of productions.

CUEING CONCEPTS

Let's dig a bit deeper into some of the examples above to explore a couple other concepts. Theatrical-style live performances are generally run in an event-based way, with a human "stage manager" in charge of running the show, and they use a headset system to communicate with individual system operators in each department (lighting, sound, machinery, video, and so on). In the classical model, "Warnings" are given for each cue; for example, "standby sound cue 13." If ready, the sound operator then replies, "sound standing by." At the appropriate point in the production, the stage manager initiates the cue by saying, "sound cue 13 go." The sound operator "takes" the cue, generally by pressing a button on a computer controller, or executing some other action. When the cue is completed, the sound operator replies, "sound complete". Of course, the "standing by" and "complete" steps are often omitted, and this model of control is often applied to many kinds of nontheatrical shows as well. Alternatively, operators may execute many of their own cues, and then take cues from the stage manager only when larger, interdepartmental coordination is necessary. If you're experienced in the business, you may be wondering why we're going over some very basic show calling procedures. But keep in mind that many show control systems implement these kinds of procedures in control systems, so we may have to model those ideas with digital data over networks. Let's look at a few more control systems concepts in that context.

Open and Closed Loop

In the example above, if the stage manager (or show control system) calls "go" on a cue, and the operator (or subsystem controller) responds "cue complete", then we can think of that system as a **closed loop** system, where a command is issued and confirmation is returned. Closed-loop systems—whether human or computer run—are generally a bit more complex than open-loop systems to implement, but generally more fault tolerant.

Alternatively, if the "standing by" and "cue complete" feedback phases of the example above are omitted, and the stage manager just issues a "standby" and then a "go" only; we could consider that system to be an **open loop** system, where a command is sent but no confirmation is returned. Open loop control systems are generally a lot simpler to implement. Keep in mind, too, that loops can be closed in many ways. A human stage manager could call a "go" command to a lighting operator, but the lighting op doesn't need to say, "complete" back over the headset, since the stage manager closes the loop visually—they can see the lights changing on stage.

System Flexibility

A tremendous amount of inherent intelligence and flexibility exists in the human-controlled, event-based system of a stage manager and operators. In the example above, the explosion effect is triggered on the event—the performer's line—regardless of when in time they say it, where they are standing, what day of the week it is, or the phase of the moon. However, if the performer is having a bad night, forgets their line, and simply runs screaming off the stage without saying anything, the stage manager can improvise and still trigger the explosion effects by giving the go commands. This type of intelligence and flexibility is complicated to build into automated systems. For example, if a voice-recognition system was designed to trigger the effects off the actor's "Run for your life!" line and the performer forgot to say it, the system could get hopelessly confused without all kinds of backup contingencies and so on, each of which adds complexity and additional failure possibilities to the system.

Absolute and Relative

Let's explore another pair of concepts. A lighting board operator taking their cue could execute it in a couple different ways. In the first case, for a theatrical (linear) show where the same show is performed in the same sequence each night, the operator might just press a "go" button on their console that executes the next cue in the cue list. Alternatively, if they are running a non-linear show, they might have to press a particular button (or fader) that is related to a particular cue. From a control system perspective, we could refer to that first "next cue go" operation as **relative**, while the one that is specifically "cue 13 go" would be **absolute**.

In a relative system, we only know where something is *in relation to the things before and after it*. A control signal sent to a lighting controller containing the message "next cue go" would be relative. The controller knows its *current* cue, so on receipt of this relative control message, it simply advances to the next cue in its list. However, if one control command in the middle of a series of commands is lost, the target system doesn't know about the error, so it can get out of whack and end up running behind the other elements to which it is connected. For example, let's say there is a series of cues such as 11, 12, 13, and 14, and these are triggered using the relative "next cue go" command sent for each cue change. We're sitting in cue 11, and then we want to go to cue 12, so the controller sends out a "next cue go" and the target device advances its cue. Then some VIP backstage kicks out a plug from the back of the console, and no one notices. The "next cue go" for cue 13 is sent, but never arrives. A technician sees the unplugged cable and restores it in time for cue 14, but, when this "next cue go" is sent, the receiving device, which is still sitting in cue 12, now advances to cue 13 when it should be in 14. The system is now messed up until someone corrects it; building that kind of intelligence into an automated system can be incredibly complicated, so data integrity is important in relative systems because if data is lost, the controlled system will be in error. But relative control approaches do have their place and are widely used—they are easy to implement and take up less resources because they carry or transmit less information.

A control message of "cue 13 go" is absolute—the control data contains all the information necessary to place the cue in the show. If one absolute message in a series is lost, the system would simply be off track until the next absolute cue is received; well-designed absolute systems will eventually realign themselves. Let's look at our cue series example from above: cues 11 and 12 are received and then 13 is lost due to the drunken VIP. When cue 14 is sent, the receiving device jumps from 12 to 14. Cue 13 was lost, but restarting with cue 14, the system can now operate correctly. Absolute systems are generally more robust and can recover from data corruption, and, therefore, are generally preferable, all else being equal. However, to generalize, they can be more complicated (and therefore more expensive) to implement than relative systems.

Absolute and Relative in Time Based Cue Lists

Let's explore these concepts a bit further in the concept of time based cue lists, where we have a sequence of cues we want to run in the same order each performance. First, here's an approach where we program into our system the *relative* time each cue should wait after the previous cue before triggering:

Cue	Action	Wait Time Until Next Cue
1	Houselights Out	00:00:05
2	Music 1 Starts	00:00:04
3	Stage Lights Up	00:00:07
4	Video Start	00:00:44
5	Sound FX 1	00:00:10
6	Sound FX 2	00:01:03
7	Video Pause	00:00:02
8	Music 1 Cuts Out	00:00:45
9	Music 2 Starts	00:00:10
10	Lights Out	

This approach works fine, and is easy to program. However, the nature of show business is change; what happens if, after you have this programmed, the director wants to move cue 3 four seconds later? You would simply change the wait time after cue 3 from 00:00:04 to 00:00:08. The problem is that now, since we increased a single wait time by four seconds, all subsequent cues will be four seconds late. To fix this, we now have go and subtract that four seconds off the wait time for cue 4 from 00:00:07 down to 00:00:03. Here's an excerpt of the update, with the changes shown in bold:

Cue	Action	Wait Time Until Next Cue
1	Houselights Out	00:00:05
2	Music 1 Starts	**00:00:08**
3	Stage Lights Up	**00:00:03**
4	Video Start	00:00:44

A more flexible solution would be to use a system that times out the cues using an absolute technique; here's the same initial cue sequence from above, but with the times shown as absolute trigger times, rather than relative wait times:

Time	Cue	Action
00:00:00	1	Houselights Out
00:00:05	2	Music 1 Starts
00:00:09	3	Stage Lights Up
00:00:16	4	Video Start
00:01:00	5	Sound FX 1
00:01:10	6	Sound FX 2
00:02:13	7	Video Pause
00:02:15	8	Music 1 Cuts Out
00:03:00	9	Music 2 Starts
00:03:10	10	Lights Out

Now, when the director wants to delay Q3 four seconds, we simply change the trigger time for cue 3 from 00:00:09 to 00:00:13. Because all the cues in the list have absolute times, no other cue's trigger times will be affected:

Time	Cue	Action
00:00:00	1	Houselights Out
00:00:05	2	Music 1 Starts
00:00:13	3	Stage Lights Up
00:00:16	4	Video Start

SHOW CONTROL SYSTEM ARCHITECTURES

For a control system to work, all the connected devices have to agree on a management scheme to dictate how the devices communicate with one other and which one is in charge at any given time. In this section, we will introduce several basic control architectures. Keep in mind, of course, that some systems are hybrids of the various types or incorporate various aspects of multiple architectures.

Primary-Secondary, Leader-Follower

In a **primary-secondary** system, one device has direct, unilateral control over another. Other terms for this approach include "main-secondary", "controller-responder", and "conductor-follower"[8]. These kinds of systems can take a variety of forms, but generally gives one device leads one or more downstream devices or systems. For example, a video playback system could send a time code signal out to an audio system that will then follow it. When the video system runs, the audio runs. While it might be possible to run the audio separately, in this structure there is no way to have the audio system force video to follow it. In another example, a show controller might send a cue to a machinery control system to move a platform 10 feet across a stage. The machinery system can either run the cue or not (perhaps due to safety concerns), but in this structure, it has no ability to tell the show controller that it's decided to only run to eight feet tonight.

A lower level variation of this primary-secondary system would be the command-response approach, where a controller asks a simple device for a response. For example, let's say we have a show control system that needs to know when a room in an attraction is occupied. The sensor system isn't able to tell the show control system what to do or what its state is; it just responds when asked with

---

8   Historically, in less enlightened times, "primary-secondary" systems were referred to as "master-slave" systems

the current state either: "The room is occupied" or "The room is empty." The only way for the show controller to know that someone has entered the room is to repeatedly ask the sensor, "What is your state right now?, What is your state right now?, What is your state right now?" When empty, the sensor responds, "empty", "empty", "empty", and then after someone enters the room, to the next poll it will respond, "occupied." This repeated questioning process is called **polling**.

Peer-to-Peer

When relatively powerful individual controllers are connected, this is often called a **peer-to-peer** system, where each has equal access to—or even control over—other devices in a system. This control structure allows for powerful and sophisticated systems, but also can be more complicated to design and troubleshoot, since contention issues—who gets control over the system at any given time—must be resolved by the system designers.

Hybrid

Since these architectures are often implemented using a network or other approach allowing easy communication, it's possible to switch architectures during a show or even between cues. For example, in one part of the show, the audio system could send out a time code that all the other systems follow; in another part of the show, the lighting system could send a signal to trigger a sound cue along with a lighting cue. In still another part of the show, the scenic automation system could export out its position for the video system to dynamically track.

Centralized or Distributed Systems

This next concept we're going to explore is related to the primary-secondary and peer-to-peer ideas discussed above, but slightly different; here, we are looking lower into the system and how it's implemented. In a centralized system, control functions are run from a single machine or system, typically in a central location (i.e., an equipment closet or a control booth).

For example, a single show control system might control all the production elements on an animatronic show.

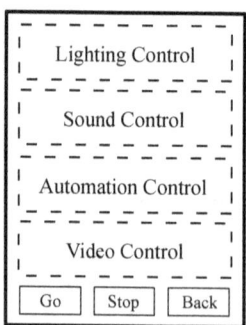

14 • CHAPTER 1 INTRODUCTION

This approach can be cheaper because less control hardware is required. However, on a show scale, with centralized control, difficulty can arise during the technical rehearsal and programming process, since many departments will want to make changes simultaneously, thus creating a bottleneck. In addition, centralized, unified control systems can end up being functional for all departments but optimized for none.

A distributed show control system is generally made up of intelligent subsystems, each optimized for its task and connected together into a larger system. Each subsystem decides, or is told by another system, when and how to execute its tasks. For example, a show control system might communicate over a network with a number of dedicated production controllers

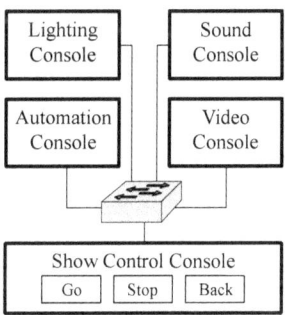

A distributed approach is often desirable, since a sound console's interface and operation can be optimized for sound, a machinery console for machinery, lighting for lighting, and so on, and multiple departments can program at the same time. This structure also has built in redundancy (we'll talk more about that in System Design Principle "2. The Show Must Go On" on page 53), and for all these reasons—all else being equal—I generally prefer distributed systems.

Other Structures

There are several other related architectures that you might encounter either as a high-level show control architecture, or, more likely in a connected systems or protocol. One type is the **client-server** model, often found in computer systems, which is a distributed structure that connects providers of some sort of resource or service, called **servers**, with **clients** that are requesting that service. These systems usually connect over a network. Similar structures include the **producer-consumer** approach, where one system produces data or signals that can be received by many others simultaneously, or the related **publish-subscribe** structure.

ENTERTAINMENT CONTROL DISCIPLINES AND DEVICES

Let's take a first look at the kinds of devices and systems that might be connected with a show control system, organized by department. Many protocols and methods developed in one department have found use in others; so there is some duplication here. And of course, this section can't be exhaustive; instead I'm just giving you some ideas of the kinds of ways that typical systems can be connected together.

Lighting

Generally lighting control systems can be broken down into a control console or "desk", and a number of luminaires of various types. DMX512 (page 39) is the primary point to point control protocol for console to device communication; control signals can also be carried over a network via sACN or Art-Net (page 40). In small systems, the show control system might use DMX or a related control method to control a few lighting fixtures directly (think houselights at a museum). In larger systems, connecting to the control console is more likely. This might be done over a network using a proprietary connection method (page 24), via another network protocol like OSC (page 25) by older point to point connection methods like MIDI Show Control (page 29), or the console might be set up to chase time code (page 75). For simple applications, it's even possible that aspects of the system might be triggered by simple on/off contact closures (page 41).

Audio

Audio plays a critical role in many productions: it can be used to amplify a performer's voice so that they can be heard in venues larger than would otherwise be possible; it can connect the audience emotionally to musical performances; or it can set an aural scene through the use of sound effects. There are a lot of internal networks and control inside audio for media distribution; typically from the show control aspect we're more concerned with connecting to control and playback devices. This might be done over a network using a proprietary connection method (page 24), via another network protocol like OSC (page 25) by older point to point connection methods like MIDI (page 27) or MIDI Show Control (page 29), or a console or playback device might be set up to chase time code (page 75). For simple applications, it's even possible that aspects of the system might be triggered by simple on/off contact closures (page 41).

Video

Dynamic imagery is an important part of live shows, and cost-effective digital video and affordable display technology has brought imagery into even the smallest of live shows in a way hardly imaginable even a generation ago. Many devices in a video system might be connected for show control applications, including video switchers, servers and projectors. This might be done over a network using a proprietary connection method (page 24), via another network protocol like OSC (page 25), by older point to point connection methods like MIDI (page 27) or MIDI Show Control (page 29), or a console or playback device might be set up to chase time code (page 75), or simple on/off contact closures (page 41). Some video servers intended for the live show market might even be controlled via DMX512 (page 39) or one of its network variants.

Lasers

Laser systems provide spectacular effects for a wide variety of productions, from city-scale light shows to concerts and corporate events. Laser projectors might be controlled by laser control software using a network based, proprietary connection method (page 24), or take DMX512 (page 39), Art-Net (page 40) or even another network protocol like OSC (page 25). Simple interfacing might take place through on/off contact closures (page 41).

Stage Machinery

This discipline covers a broad range of automation control systems that are used in scenic-related applications on shows, including stage elevators, scenic automation systems, automated rigging systems, mechanized props, or theme park show-action equipment. Since such systems offer control over motion, they are often called motion control systems, a term borrowed from the industrial automation industry. Stage machinery, obviously, is inherently dangerous and any system designer should always take safety into account and consult an expert before attempting to do anything in this area. But many scenic automation systems are built on networks and therefore can be remotely controlled (using safety features, of course) This might be done over a network using a proprietary connection method (page 24), or a controller might be set up to chase time code (page 75), or things could be interfaced using simple on/off contact closures (page 41). It's also possible for many scenic automation systems to output their position (page 46) for following by other systems.

Animatronics

In animatronics, or "character animation," machines designed to look like living characters perform to prerecorded sound or video tracks, live puppeteers, or even interact with guests or audience members. Animatronics incorporates aspects of stage machinery, audio, robotics, and computer control systems. Animatronic control systems might chase or generate time code (page 75), communicate via a network using a proprietary connection method (page 24), simple on/off contact closures (page 41) or even DMX512 (page 39).

Fog, Smoke, Fire, and Water

This category contains a variety of production elements, some usually found only in theme parks: fog, flammable gas, water fountains, and such. Most of these systems (other than fog) are generally purpose-built, with a custom interface to other show systems. Some of these systems can involve serious safety considerations which should always be considered and experts should be consulted. Fog machines are often controllable using DMX512 (page 39), and more sophisticated systems could communicate via a network using a proprietary connection method (page 24), chase or generate time code (page 75), or interface through simple on/off contact closures (page 41)

Pyrotechnics

Here we are talking about chemical pyrotechnics: flash pots, sparkle devices, concussion mortars—things that go boom (pyro substitutes such as air cannons would go into other categories, depending on their construction). Pyrotechnic systems have obvious safety considerations that must always be considered, but controllers might chase or generate time code (page 75), interface with simple on/off contact closures (page 41) or communicate via a network using a proprietary connection method (page 24).

Building Systems

In permanently-installed show control systems, building automation systems such as HVAC, fire and security alarms, elevators, and related systems might be monitored, interfaced, or even controlled. Generally this would be via a network using a proprietary connection method (page 24), or for simple applications with on/off contact closures (page 41).

Other Show Controllers

In some complex systems, multiple show control systems may be required. Generally the most straightforward way to interface these systems would be over a network via one of the many available protocols.

MOVING ON

Now that we've introduced the topic of show control, we will progress through several chapters expanding on all of these ideas. Next up in Chapter 2, "Connecting Devices and Systems" on page 21, we will explore a variety of ways that devices and systems can be connected. Then, in Chapter 3 "Show Control System Design Process" on page 58 I will introduce my design process, and then finally in Chapter 4 "Example Show Control Systems" on page 65 we will apply that design process to realistic examples.

Chapter 2

CONNECTING DEVICES AND SYSTEMS

As we've discussed, show control is the practice of connecting two or more entertainment control systems together, and since the array of possible connected devices is so vast, the process of connecting things together can take many forms. Show control systems can have direct inputs from a variety of devices, connect to other systems using various forms of data communication or—most commonly—connect over a network using a variety of protocols.

Because the possibilities are so numerous, this section can not be exhaustive or comprehensive. Instead, it's intended as a survey of possibilities; the show control system designer can dig further into many of these connection approaches as needed. I will go a bit deeper here into methods that are more specifically show control-oriented; but before we do that let's take a look at the ways digital communications signals are sent on a show.

DATA COMMUNICATIONS

Show control systems deal primarily with digital information; that digital information—representing everything from cue numbers to lighting levels—has to be converted to or modeled in a digital format, and then communicated to the show control systems in some way. There are a few ways to send this data around a show control system.

Legacy Point to Point Standards
Many 1980s vintage standards like DMX512 (page 39) and MIDI (page 27) are based on older **point-to-point serial** interfacing standards. These interfaces carry digital data from one device to another, or perhaps a few devices. The problem with point-to-point approaches is that they do not scale up well for larger systems, since they require direct connections from each device to every other device in the system with which communications is desired. Some of the older, common point-to-point standards included RS-232, RS-422, and RS-485; lots of information is available on those interfaces online. Most protocols designed since the 1990s are, of course, based on networks.

Networking

As we discussed on page 1, instead of individual, point-to-point connections, a network uses a common physical infrastructure that still allows each device to have access to all the others. Connections between devices then simply become *virtual* pathways, and the resulting system can be dramatically cheaper, more powerful, simpler overall, and far easier to manage. For a thorough introduction to networks on shows, please see the companion book to this one, *Introduction to Show Networking*[1].

Other Connection Methods

While networks and older serial point-to-point connection methods are the most widely used communication methods for most of show control, there are some other ways to communicate data that you might encounter, especially in short haul applications. **Universal Serial Bus** (**USB**) comes in a wide array of flavors and of course is widely used for computer peripherals, and can be used for some show control communications. **Zigbee** is used to create small wireless networks and is commonly found in home automation and similar applications, and since we might connect to these systems you might find applications for it. **Bluetooth Low Energy** (**BLE**) is a descendant of Bluetooth, widely used for connecting devices to phones and related applications. BLE allows small devices to run for a long time on a battery.

Control Protocol Approaches

To exchange information, computers use a **protocol**, and since show control systems are built from computers, we too use protocols. What is a protocol? One definition from the *American Heritage* dictionary[2] that makes sense for our purposes is, "A standard procedure for regulating data transmission between computers." There are many protocols in use in show control, and the low-level details can be mind-numbing. Fortunately, our equipment and software programmers and designers have packaged most of that complexity into our products, so the average end user doesn't need to have their mind numbed in order to use these powerful digital communications methods. However, from a conceptual level, there are two general ways that show control protocols typically digitally communicate with devices, and I've named them here **preset/command** and **repetitive/continuous**.

---

1 Details on how to get the companion networking book here: https://www.controlgeek.net/bookinfo

2 https://www.ahdictionary.com/word/search.html?q=protocol

Preset/Command

In preset or command-based control, the target (controlled) device needs enough intelligence to be able to activate some sort of previously recorded preset that can be recalled when the command is received, or it needs to understand enough to be able to interpret or act on a command immediately. The device then does something or issues a low-level command to another device. For example, a command may be something like "go cue 66," which would recall a preset scene numbered 66 in a sound console. Alternatively, commands can also take on a more detailed, non preset-based form; for example, "move to 43 feet, with an acceleration time of one second and a deceleration time of two seconds." To a control system, however, both of these approaches look pretty similar: a relatively short message triggering a potentially complex action. This structure requires a relatively significant amount of processing power in the target device, which is readily available on higher-level controllers (e.g., lighting consoles). For lower-level devices, which may not have as much processing horsepower, a "continuous" control approach may be taken, with the control data sent repetitively.

Repetitive/Continuous

"Dumb" target devices require less local intelligence. They don't understand anything about their larger context—they simply react to the data they are receiving *right now*. In a lighting system, for example, there may be many fixtures, each of which needs to have its individual lighting levels set, while a lighting control console provides the global cue storage and interface for the system operator (and show control system). When the widely-used DMX protocol (page 39) was developed, moving and LED lights didn't even exist; it was cost prohibitive (and overly complex) to engineer each dimmer to provide the ability to store, recall, and process cues. Instead, each dimmer is designed to simply set its connected light to the desired level, based on the data it is presented at that instant.[3] If the dimmer needs to fade from a level of 50% to 60%, then the target level is transmitted at 50%, then 51%, then 52% and so on up to 60%, with the dimmer immediately changing its level based on each piece of new incoming, repetitively sent data. Alternatively, if the dimmer is sitting "off" at a level of zero, then the zero data value is simply sent repeatedly over and over and over. This same approach might be used in a networked system where speed is of the essence; for example in a moving scenic platform exporting its position out so the video system can track it. Here, if a positional update is lost, it's more important to get the next update (which will overwrite the position anyway) out as fast as possible, rather than renegotiating to resend the lost information. In this case, a slight glitch might happen, but that's a

---

3 You could look at this message as a "command," but for our purposes, commands are sent only once, not over and over and over again.

small price to pay to get the tracking as fast and accurate as possible.

There are advantages to this approach: Controlled devices can be made very simply and cheaply and, if control data is lost or corrupted, the corrupt data will simply be overwritten by the next update. There are also drawbacks, of course; communications bandwidth and processing power is wasted in sending out data on a repetitive/continuous basis since—even if nothing is changing—data has to be sent out over and over and over again. But we have lots of capacity on networks so this isn't as much of a concern as it once was.

With those basics in mind, let's move onto a number of different protocols used in show control applications, sorted in no particular order but more or less by my perception of how widely used each is in the industry.

CUSTOM NETWORK-BASED PROTOCOLS

With Ethernet as our common digital highway for digital show communications, perhaps the most widely used way of controlling and connecting equipment is using a custom protocol designed by the manufacturer of the gear you are controlling, or even a protocol you write yourself. This can take the form of proprietary low-level binary data representing positional information, or manufacturer-defined commands such as "Go Cue 59". Alternatively, a text-based language used by the target system can be exposed onto the network, so that the show control system is effectively typing on the device. For example, on a lighting console a syntax of `goto cue 27.8 exec 1.2` followed by a carriage return (enter key) could load cue 27.8 on page 1, executor 2.

ASCII

In this kind of protocol, alphanumeric characters are generally encoded using the American Standard Code for Information Interchange (**ASCII**) which is part of the UTF-8 international standard. In *Introduction to Show Networking*[4], I describe ASCII this way:

> ASCII (pronounced ASS-kee) was standardized in the early 1960s, and is basically a grown-up version of a communication game you may have played as a child: substituting numbers for letters of the alphabet in order to send coded messages. ... ASCII was, and UTF-8 is now, one of the most widely-used standards in computing and networking, and many

---

4 Page 8 of the first edition.

other standards reference or use them. UTF-8 and Unicode are backwards compatible, and in basic control systems we're likely to be using pretty simple characters, so most people call just this standard "ASCII"[5].

Network Details

In order to get these messages into the device over the network, the Internet Protocol (IP) address, the transport protocol (typically Transmission Control Protocol (TCP) or User Datagram Protocol (UDP)), and the port number all need to be known and configured. All that and more is covered in my *Introduction to Show Networking* book[6].

Because each of these protocols is specific to a particular device, there's not much I can cover here in general. However, I do have many detailed examples of this kind of communication on my blog[7].

OPEN SOUND CONTROL (OSC)

Another widely used, network-based framework for communication is Open Sound Control (**OSC**), a 1990s-era open protocol for message-based communication between computers, synthesizers, samplers, gestural controllers, and other similar devices. OSC is also used in show control to connect a wide variety of devices, everything from lighting controllers to sound playback systems. OSC doesn't actually contain definitions of any sort of functionality that a user might use—no commands like "Go" or "Play". Instead, OSC defines a structure for an "address space" and then lets the system or product designers decide on all the functionality that makes up that address space. In some ways, OSC is sort of analogous to a spreadsheet, in that it doesn't describe or standardize a user's data, but instead gives system designers a structured way in which to exchange that data.

Clients and Servers

OSC was designed to run on a wide variety of communications systems, but of course is mostly run over networks. Any device in OSC that sends OSC packets is a client; a device that receives packets is a server. Clients in OSC can only connect to one server at a time, but servers can receive packets from many clients. For example, sound generation software running on a computer (server) could be

---

5 There are many websites available that show the ASCII characters and related numbers.

6 Info on the book here: https://www.controlgeek.net/bookinfo

7 https://www.controlgeek.net/

controlled simultaneously by two different kinds of controllers (clients).

Addresses

The OSC server has an address space that gives a tree structure to a number of "containers," which contain "methods," that actually accept and act on the "arguments," or data. Each method and container has a name, which is simply a string of ASCII characters. For example, a mixer (OSC server) could have several channels, each with a volume control method called "Volume," another method called "Pan," and another for "mute." It might accept a level as its volume method an argument for level "50" at the address /Mixer/Channel_2/Volume:

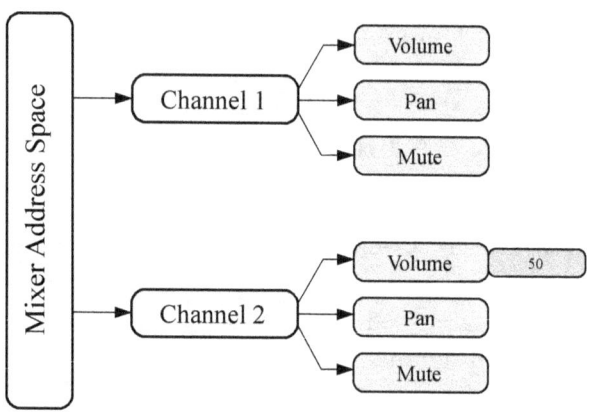

Examples

Each manufacturer of controlled equipment makes their own specific protocol, and the syntax can be simple or complicated. But here's a couple examples of simple, realistic implementations.

To fire cue 1 on cue list 1 on a lighting console might send something like:

/cue/1/1/fire

Or, to trigger cue 123 on a sound console, you might send something like:

/cue/123/go

Keep in mind that because this is a protocol which often runs over networks, the actual targeting and delivery mechanism of the message is outside the protocol itself (unlike some of the older protocols like MIDI, page 27), so the messages can be very simple. Additionally, OSC can communicate bidirectionally, and can be used to query parameters of a connected unit, which can return all kinds of information dynamically.

OSC in Show Control

In the 2010s, OSC started gaining broader acceptance in our market beyond traditional musical and sound applications. Fortunately, in most implementations in our market, the user is isolated from the low-level communication details of OSC; the user simply has to make sure the devices can communicate (typically this just involves configuring a network), and then make sure the right OSC messages are assigned to the right parameters of their controlling device and target.

MUSICAL INSTRUMENT DIGITAL INTERFACE (MIDI)

The Musical Instrument Digital Interface (**MIDI**) was originally designed for synthesizer interconnection in the early 1980s when, with the explosive growth of sophisticated keyboard synthesizers, musicians began to want their keyboards linked to simplify complex studio and live-performance setups. In its most basic configuration, MIDI allows instruments to be connected in a simple primary/secondary relationship, transmitting one way commands between the two.

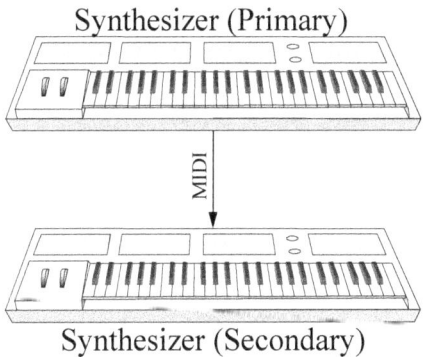

Like DMX (page 39), MIDI was developed before layered, network-based approaches became commonplace in our industry, and so MIDI defines the connector (5 pin DIN), the electrical interconnection interface, and the command set. No audio is transmitted over a MIDI line; only control information representing musical events is sent. MIDI is open loop and very simple, but can be useful for simple show control or musical-style applications.

MIDI Messages

MIDI contains a wide array of messages designed for musical applications, and was designed to transmit over 16 "channels". Because it was designed to run on a simple, point to point electrical interface, the channel number is included in the message so that a receiving device can look at the channel number and decide whether or not it should listen to messages on that channel, depending on how it is configured. If the incoming message matches the channel configured into the receiving device, then the device can decide whether it understands the incoming message and then act on it if appropriate. Common MIDI messages include the following.

Common MIDI Messages
Note Off
Note On
Control Change
Program Change
Channel Pressure
Pitch Bend Change

In addition to the channel, most message types contain some sort of data. For example, Note On messages contains a note number indicating the associated keyboard key, and another number representing the "velocity" at which the key is struck. In this way, different keyboard sounds (and intensities) can be modeled. Other messages like Program Change can indicate a limited number of program numbers (think different sound patches on a synthesizer), and messages like Pitch Bend send out changes as a pitch bend controller is moved.

MIDI Hardware

MIDI was designed to run over special cables no more than 50 feet long[8]. Because interfacing hardware and computer horsepower was expensive in those days, it was common for a single hardware MIDI output to be cheaply split and the same electrical signal—and therefore all the messages transmitted over that signal—was sent to lots of different pieces of equipment, with all messages broadcast to all downstream devices. Over time, it became easy to separate the MIDI message data from the original MIDI physical connection and instead run the messages over USB for short-distance computer connections, or over a network for longer distances or more sophisticated applications.

---

8 Because of the electrical characteristics of the MIDI signal, it actually can be run quite a bit farther than 50 feet, depending on the cable quality. It's commonly run over audio snakes or even Cat 5e cable for long distances; test your setup before counting on this though.

MIDI Applications in Show Control

"Musical" MIDI can be useful for musical-style applications on shows, where the messages map to functions on a show (like triggering sound effects). Additionally, it can be used for simple show control applications like representing a contact closure (page 41) on/off, even for things like triggering video clips. The designers left room for expansion in the specification, through "System Exclusive" messages. These messages could be based around an ID assigned to a specific manufacturer; some IDs were reserved for other non-manufacturer-specific applications, and that allowed development of things like MIDI Show Control (page 29) and MIDI Time Code (page 36).

MIDI SHOW CONTROL (MSC)

MIDI Show Control (**MSC**) is an extension to MIDI made in the early 1990s for show control applications. It contains commands like "Go", "Stop", "Resume" and was designed to be a simple and universal standard, with the ability to control everything and anything on a show.

Targeting Devices

As we discussed above, MIDI—and by extension MSC—was developed in an era where, due to cost limitations, it was common for a single electrical MIDI signal encoding the MSC data to be sent to several or even all of the control devices on the show. For this reason, it was designed in a way to allow a receiving device to know whether or not a particular message was intended for it. While musical MIDI contains 16 channels, this wasn't enough for the show environment, so to allow specific device targeting over this broadcast protocol, MSC includes a "Command Format," and a "Device ID".

In general terms, the Command Format indicates what type of equipment is intended to be addressed by a particular MSC message. A large variety of Command Formats were defined in the standard in general categories:

MSC General Command Formats
Lighting
Sound
Machinery
Video
Projection
Process Control
Pyro
All-Types

The final "All-Types" Command Format is intended for all devices listening to a particular electrical MIDI signal.

Inside the General Command Formats are a number of kinds of specific equipment. For example, inside the Sound general category are things like CD players, audio tape machines, and EEPROM players. Because many of the specific devices recognized in the original standard are obsolete, most MSC is addressed to the general category Command Formats.

Given that MSC messages may be broadcast to a wide variety of devices simultaneously, it's possible you might have a situation where something like two lighting consoles are connected, but need to be communicated with individually. For this reason, the Device ID is included, allowing 112 individual devices to be addressed, 15 groups, or every receiving device in the Command Format via the "All Call". Many control devices allow the user to configure the Device ID, although this flexibility is not specifically called for in the standard.

Cue Data

As one of the few control standards designed specifically for the live show control market, the cue number is a vital part of MSC. Cues are addressed via Cue Number and optional Cue List and Cue Path. Cue List indicates to the receiver in which of the currently "open" Cue Lists the Cue Number resides. These meanings aren't set in stone, though; lighting equipment, for example, might use the Cue List number to indicate which playback control or fader is being used to execute a particular cue. So, if a lighting console had eight faders, Cue List value of 1 could be used to tell the console to execute the cue on fader 1; Cue List 2 could denote fader 2, and so on. Cue Path is rarely used, but was intended to indicate from which of the "available media" the Cue List should be pulled.

Some Examples

So some typical MIDI Show Control messages would include:

```
Lighting-General, ID 2:

Go Cue 72.5
```

This would be intended to tell a lighting console which has been configured to receive on Device ID 2 to run Cue 72.5.

```
Lighting-General, ID 2:

Go Cue 72.5, Cue List 7
```

This would be intended to tell the same lighting console configured to run Cue 72.5 on fader 7.

Alternatively, the next cue which is standing by could be told to go:

```
Lighting-General, ID 2:

Go
```

This would be useful for a simple remote Go button.

And there are a huge variety of possible commands. For example, the following:

```
Sound-General, ID 13:

Go Cue 66, Cue List 4
```

would be intended to tell the sound controller configured to receive on Device ID 13 to run Cue 66 out of Cue List 4, and so on.

Typically on show control software the MSC message can be just selected from a pull down list by the user, hiding the low-level details when building cues.

Limitations of MIDI Show Control

MSC found some common applications: controlling lighting consoles, triggering computerized sound systems, and so on. However, MSC was designed to be a universal standard, with the ability to control everything and anything on a show, and this universal command approach, in my opinion[9], was more desirable in the 1990s, when we did not have the universal data transport facilities offered by Ethernet and IP. Additionally MSC is open loop, so there no feedback or confirmation of any kind is required for the completion of any action. When a controller sends a message out, it has no idea if the target device even exists. And as for MSC

---

9 I have my reasoning for this on my blog, starting here: https://www.controlgeek.net/blog/2008/10/21/in-praise-of-custom-ip-based-entertainment-control-part-1.html

applications that could be potentially life threatening, the standard clearly states that MSC is intended only to "signal what is desired if all conditions are acceptable and ideal for safe performance". Safety is the responsibility of downstream systems and trained show personnel.[10] Finally, building MSC on MIDI (which certainly made sense at the time) imposed some limitations[11] for more complex applications, and I prefer approaches like sending simple manufacturer protocols over a network or OSC. But MIDI—and MSC by extension—offers a reliable, simple, limited connection method useful for simple control.

TIME CODE

Time code (TC) has its origins in videotape editing going back to the 1960s[12], and was designed to allow analog, linear media (audio and video) recordings (on tape back in the day) to be synchronized. It breaks time down into hours, minutes, seconds, and frames; up to 24 hours worth of frames can be encoded, although it's important to remember that the TC signal is its *own clock*, and does not necessarily match real time of day. One discrete TC "address" exists for every frame on an encoded media—a typical address might be `13:02:57:18`, or thirteen hours[13] two minutes, fifty seven seconds, 18 frames.

Time code frames are absolute, and the signal was designed to be recorded on a wide variety of media to designed to synchronize one or more secondary devices to a primary time code source. In its simplest form, when the primary rolls forward, the time code frames start incrementing, and the secondary sees the changes and syncs to and **chase**s the primary.

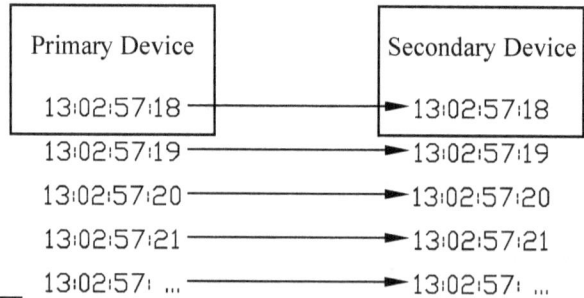

10 MIDI Show Control version 1.1 was developed, added some additional commands to enhance safety, but never caught on.

11 The biggest limitation of MIDI is that it is structurally limited to only transmitting 7-bit integers.

12 See my blog if you're interested in the history: `https://www.controlgeek.net/blog/2023-2-17-a-bit-of-smpte-time-code-history`

13 1pm in non 24 hour format, although usually in live show time code we would just say "hour 13".

32 • CHAPTER 2 CONNECTING DEVICES AND SYSTEMS

Time code is rarely recorded on analog media any more, but is instead either generated in real time or recorded digitally and then sent from one computer to another, and it has found uses far beyond its originally intended applications.

Show Control Time Code Applications

While time code's roots are in the film/video production and post-production worlds, we oftentimes want to do very similar things in the live entertainment industry. We may want an audio multi-track recording chase a video; or have four video servers run separate images to separate screens, all in synchronization with an audio track; or have an animatronic character synchronize tightly to a soundtrack.

In addition, we also use time code to trigger event-based systems. For instance, if we need to synchronize light cues to a sound track, we could record (or generate) TC along with the audio. Then, whenever the audio track is rolling, time code would output to a lighting console. Professional lighting consoles generally accept time code directly, and each cue can be programmed to be triggered at a precise time: Cue 18 might be programmed to execute at 07:02:59:23. Whenever the audio track rolls and the preprogrammed time code address comes along, the appropriate light cues trigger. In a similar fashion, many preset-based audio mixers, video routers, pyro and fireworks controllers, or other systems can be triggered.

Frame Rates

Since different media and countries have different framing rates, there are different types of time code to match. Monochrome American video shared the 60-Hz frequency of North American AC power mains, and because it was interlaced,[14] the framing rate is exactly 30 fps, hence 30-frame time code. In Europe and other places around the world, they based their video frame rate off their 50-Hz power line; so time code for those applications often 25 fps. There are many other flavors of time codes for other frame rates; but the most confusing of all is "drop frame".

Drop-Frame Time Code

Color video in America was standardized in the 1950s (and used for over-the-air broadcast until 2009 in the US), and was designed to be backwards compatible with monochrome TV. For reasons outside the scope of this book (regarding the encoding of the color information), traditional color video in the US runs at a frame rate of approximately 29.97 fps. If regular 30-fps TC is used to count video

---

14 Every other line is scanned first, and then the remaining lines are scanned on alternate "fields." Two fields make up one frame.

frames running at a rate of 29.97 fps, the time-code second will gain 0.03 frames per second in relation to a true second; a cumulative gain of 108 frames, or 3.6 seconds per hour of running time, will result.

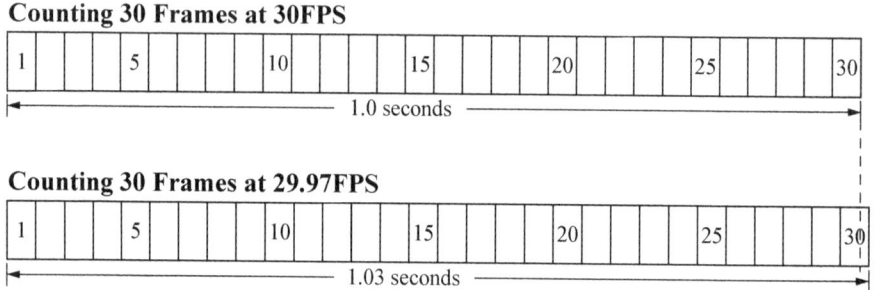

This error may be a bit difficult to comprehend, but the key is that time code receivers don't (generally) compare the frame counts of the incoming time code signal with an external clock, because the incoming TC *is a clock*. So, if we're using 30 fps TC to count American color video frame rate (NTSC) 29.97 fps frames, it will take 1.03 actual seconds to reach a count of 30, or 1 time-code second. While an error of +0.03 real seconds per time-code second doesn't seem like much, this accumulates to 3.6 seconds per hour, or 1 minute and 26.4 seconds per day. It would certainly be a problem if a light cue was off by 3.6 seconds at the end of an hour-long presentation.

For this reason, drop-frame TC was developed; it gets this seemingly strange name because the extra 108 frame counts that would accumulate over the course of an hour are simply omitted. Conceptually, this is similar to a sort of reverse "leap-year," although here we're dropping frames every minute instead of adding a day every four years. Since it is impossible to drop fractional frames, the first two frame numbers of every minute, with the exception of minutes 0, 10, 20, 30, 40, and 50 are dropped Even this method still only approximates real time; drop-frame TC still has an error of −3.6 milliseconds every hour. Even though this type of time code has its roots going back to the 1960s, these odd frame rates may still be found. But due to their innately confusing characteristics I recommend avoiding drop frame whenever possible.

There are two types of time code widely used in the show control market; SMPTE (pronounced "sim-tee") time code and MIDI Time Code (MTC).

SMPTE Time Code

There are two types of **SMPTE** time code: **Linear Time Code**[15] (**LTC**), and mostly obsolete and rarely-used Vertical Interval Time Code (VITC) pronounced "vitcee"[16]. Linear Time Code is so named because it was designed to be recorded on an audio track linearly along the edge of a videotape, not helically like the video signal. LTC can be distributed as and works similarly to an audio signal, although it is a square wave with harmonics that can quickly get out of the audible range.

SMPTE Time Code Hardware

Because of the extensive broadcast equipment market, and their historic acceptance of the SMPTE Time Code standard, there is a large variety of specialized time code generation, processing, distribution, and interfacing hardware. Many live shows simply use one of the show systems (audio, video, etc) to be the primary time source, but in some cases it's desirable to have a single, external, time code generator, which can then be started and stopped to run the show. This can be a function of show control software; stand-alone time-code generators are also available that can sync to time of day (typically via the highly accurate GPS system, or other approaches like the Network Time Protocol (NTP, page 46). This allows shows that trigger via time of day (for example, an hourly fountain show) to be precisely controlled. Another invaluable device is a time code display, which simply shows the current time code addresses as they roll by. This is a very useful device for troubleshooting and system monitoring.

LTC Distribution

SMPTE LTC is distributed using audio equipment; that means you should use good audio practices to distribute it. Use good quality cable, and keep the time code level above the noise floor and safely below the clipping level. Being close to the noise floor could add enough noise to cause problems with the signal; running it too high could cause distortion, corrupting the signal. In this vein, it's good practice to use some sort of audio distribution amplifier (DA) to split/distribute the LTC signal; this approach isolates each output from others on the same link, and ensures that each device receives a clean signal.

LTC Regeneration

LTC degenerates each time it is recorded or rerecorded, especially with (obsolete) analog systems. So, historically, it was recommended that any time you re-re-

---

15 Also sometimes called "longitudinal" time code.

16 VITC was developed in the late 1970s after videotape recorders capable of playing back high-quality still images were developed. In its traditional recorded form, LTC can only be decoded when the tape is moving, and becomes unreadable when the media is not moving.

corded LTC you "regenerate" or "jam-sync" it, cleaning up bit transitions through reshaping. This is generally not needed with digital LTC recordings, or with time code that is generated directly.

MIDI Time Code (MTC)

Since SMPTE TC is sent over analog audio interfaces, machines that have only digital inputs (e.g., computers) must sample and convert an LTC signal somehow in order to extract the time code data. Alternatively, systems can use **MIDI Time Code** (**MTC**), which breaks the SMPTE TC frames into MIDI messages and digitally transmits them directly down a MIDI line.

MTC Messages

MIDI Time Code is transmitted in two formats: full messages and quarter-frame messages. Full messages contain a complete time code address and are sent whenever a system starts or stops, or when the system moves in or out of various transport modes (i.e., from Fast Forward to Play). Quarter-frame messages transmit pieces, or "nibbles," of the current TC address, and as the name implies, they are sent out at the rate of four per TC frame. Eight quarter-frame messages are used to communicate one full TC address, so the time-code count can be fully updated only every other frame. However, a receiving system can still *sync* four times per TC frame, by examining the timing of the quarter-frame messages.

MIDI Time Code Specific Recommendations

MIDI Time code is actually theoretically capable of *higher* resolution than SMPTE, because it uses four quarter-frame messages per single SMPTE frame. At a typical frame rate of 30 fps, 120 MTC messages would be sent out each second. However, some MTC messages—especially with old school, hardware-based MIDI distribution—might be delayed in a MIDI merger or other processing device due to heavy traffic. So, if timing is critical in a particular application, a dedicated MIDI link should be run directly from the generating device to the receiver, with no other MIDI traffic allowed on that segment. If the MIDI link is simply a piece of cable with no other traffic, it will have a predictable latency.

Practical Time Code For Live Shows

There are a number of issues to keep in mind when dealing with time code on live shows.

Time Code Hours/Pre-Roll

Shows that run 24 hours are rare,[17] so it is typically more convenient for us to use the time code hours for other purposes. For instance, time code hour 1 might be used for show one, hour 2 for show two, and so on.

Keep in mind that it's always a good idea to leave some time before the show start for pre-roll, to make sure everything is synchronized; this is why shows often start on an even hour like 01:00:00:00 with pre-roll (10 seconds in this case) starting from something like 00:59:50:00.

Agree on a Frame Rate

The most important thing in any time code application is to be sure that the show's frame rate is chosen and agreed upon well in advance, everything is double-checked, and time is left for testing the assembled system.

Use LTC rather than MTC

Practices to distribute LTC over an audio line are well known and robust. MTC—especially with old school, hardware-based MIDI distribution—can work perfectly well, or might have some subtle issues that are hard to diagnose. For this reason—all else being equal—I would recommend using LTC.

Stay Away From Drop frame Time Code If You Can

Drop frame time code was developed to solve a problem in resolving now-obsolete, analog video to a time of day clock. If we are running a 20 minute show that is not related to the time of day (we start our show at hour 2 when the house is ready around 8pm, for example), then we don't really care about how the time code lines up with the time of day clock; the time code clock is just a relative time trigger for us to use. So on straightforward live shows where mostly events (light cues, fireworks, etc.) are triggered from some prerecorded media, a simple 30 frame per second (or 25 in Europe), "non-drop" time code is often used, simplifying everything. Using this frame rate, over time, the time code could end up being slightly different than the time code on video running at the slightly slower 29.97 fps. But if we're just using the time code for triggering events, we can simply ignore the discrepancy because the 30fps time code will be consistent, and the cues will fire on time.

---

17 Although there was at least one such show in Times Square at the 1999/2000 New Year's Eve celebration. For that show, they had to figure out how the machines dealt with rolling over from 23:59:59:29 to 00:00:00:00.

Chasing or Not

Another feature of well-designed equipment that handles time code is the ability to somehow *disable*, or not listen to, or chase, the incoming time code stream (and ideally, this chase on/off should be remote controllable). This is extremely useful when you are trying to work on a system and someone else is doing something that causes the time code to roll—it is really annoying when you are deep into programming a cue and your console starts executing cues based on incoming TC!

Dropouts and Freewheeling

Occasionally, in the old days, due to tape wear, connection difficulties, or other problems, time code "drop outs" would occur, where the signal momentarily disappears or becomes too corrupted to decode. While this is rarely a problem any more, many devices are capable of "freewheeling" through these dropouts by simply rolling on with their own time bases until the code reappears or can be decoded properly again. This could be good, covering a momentary drop out, or could actually mask a serious problem and roll onwards when the time code had actually stopped. On any well-designed piece of equipment incorporating this feature, freewheeling should be able to be enabled or disabled, and the number of freewheeling frames should be configurable.

Figuring Out How to Stop

One of the drawbacks of time code is that, since it's just a series of frame addresses, when the incoming frames stop, the receiving device doesn't know whether this is the end of the show, or even just a drop out. So it may be necessary to include some "out of band" control signal (or human monitoring) to know for sure that the show is on or it's time to move to the next segment.

Time Code Over a Network

There are a number of ways to send time code signals over a network. As of 2023, there is no widely-used standard way to do this.

OTHER WAYS TO SYNCHRONIZE SYSTEMS

Time-based shows are very common, but of course you don't have to use time code to have a time-based show.

Running Wild

Time code was developed in the days of all-analog equipment, which could drift out of sync easily. Most computers have incredibly accurate time bases, so it's often possible to simply start two digital systems at the same time and have them run **wild** together very accurately. Of course there is no guarantee that things will stay in sync, but you might be able to get away with this approach, especially for short shows. Testing is important to verify accuracy here.

Proprietary Time Codes

Many show programs and media servers create an internal, non-standard time code clock based on the internal clock of the computer or its software operations. These time codes can sometimes be sent to other systems, or used within a proprietary system.

Click Track

A **click track** is typically an audio track used to tell musicians the tempo of a song, and often indicates when the piece begins. The track is traditionally a series of metronome clicks starting a measure or two before the band should start; hence, the name click track. Click tracks can also be things like vocal count-offs—"1 2 3 4 1 2 3 4 Go"—or drum patterns. The click track is often sent via an in-ear monitor to the drummer or the conductor, and can be used to have musicians sync with prerecorded audio or video tracks.

DIGITAL MULTIPLEX (DMX512)

DMX512 started humbly in the mid 1980s, but is the most successful and widely-used entertainment lighting control standard in the world and is used to control all kinds of gear, some not even in existence in the 1980s: moving lights, LED fixtures, and even video servers or laser projectors. I'm including it here because sometimes show control systems must control a few lights or devices directly, and the easiest way to do so is with DMX; additionally, sometimes DMX is used as a simple interface between systems.

"DMX" stands for Digital MultipleXing; "512" is the number of control "slots" available for transmission in each "universe" (when more than 512 control channels are needed, multiple universes are implemented). DMX was designed to be sent over a dedicated point-to point serial interface on five-pin XLR connectors[18]. DMX doesn't contain any routing intelligence and simply sends its data values in a simple repetitive control approach; each slot can be set to an eight-bit value of 0-255, which is often mapped by control systems into a 0-100 percentage. So to cause a light to fade up, the console might send a series of updates

---

18 You may see DMX on three-pin XLRs, but three pin connectors are not even mentioned in the standard. Five-pin was chosen to keep from cross plugging with audio and this is still best practice.

like 00,10,12,15,20,25,... etc. until the desired level is achieved. If a slot is sending "off" instructions, it will just be a repeating set of 00 updates: 00,00,00,00,00, etc. If all 512 slots are sent, then the signal will be updated about 44 times a second.

DMX data is broadcast from the console to all connected devices; it's then up to the technicians to configure the receiving devices to tell them which slot or slots to listen to. This slot arrangement is developed by each fixture's manufacturers; a simple LED wash fixture might just have a few slots allocated to red, green and blue intensities, while a moving light might use a bunch of slots to map out all the controlled functions.

DMX Over a Network

DMX was designed in the 1980s as a complete, unified standard, specifying everything from the connector to the cable type to the voltages on the pins, and from the timing of the signals to the meaning of each bit. Given DMX's limitations it's often desirable send DMX over a network; there's two ways to do this in widespread use: **Art-Net** and **Streaming ACN (sACN)**. Art-Net was one of the first developed as an easy-to-use way of transporting DMX and RDM (page 40) over closed, private Ethernet networks. sACN is a sophisticated network-centric protocol designed for the transport of DMX-like control data.

REMOTE DEVICE MANAGEMENT (RDM)

DMX dominates the world of lighting control because it is simple to use and implement, and it works well. But that simplicity comes with limitations. **Remote Device Management** (RDM) is an extension to—not a replacement for—DMX, standardizing an open, bi-directional communication approach, bringing extended functionality to a DMX infrastructure. It also allows a control console to automatically "discover" connected devices, and then configure, monitor, and manage them from the console—all without getting out a ladder or a personnel lift in order to push buttons to set the address on a moving light up on the truss, for example. RDM can also deliver error and status information, so a problem might even be able to be corrected before it becomes visible on stage. RDM uses DMX's point to point serial interface, allowing half-duplex, bi-directional communication between any connected devices on the same physical link. **RDMnet** allows the RDM functionality to be implemented on a network.

CONTACT CLOSURES FOR INPUT OR OUTPUT

One of the simplest (and most limited) ways to interface two things electrically is to use a **contact closure**, sometimes known as a "dry" contact closure (DCC) or **General Purpose Interface** (**GPI**). This type of interface is a simple, limited, and very effective way of connecting two systems, since only two states can be communicated—on or off, binary 1 or 0.[19] Simple as they are, contact closures are quite useful for many either/or, yes/no entertainment control applications, such as a "Go" button being pressed on a control console by an operator (the button can only be on or off), a sensor indicating to the control system that audience members have entered or left a room in a themed attraction, a motor turning on or off, or a thermostat indicating that a target temperature has not yet been reached or has been reached.

But looking from an electrical perspective, you can visualize a contact closure like an old-timey knife switch:.

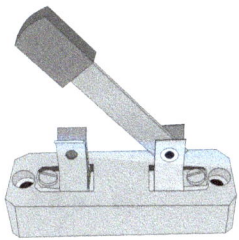

A contact closure interface may contain physical, conductive contacts. Alternatively, the on-off functionality may be provided by a solid-state (no moving parts) "switch" such as a transistor. Let's look at some contact closure input and output applications.

Contact Closure Inputs

Contact closure interfacing devices might be wired directly into a show control device; alternatively they can be connected to a remote interfacing device with their status monitored by the show control system over a network. With this simple structure, a wide variety of information can be reported into the show control system for processing.

---

19 Note that I explain binary in the *Introduction to Show Networking* book. Details here: https://www.controlgeek.net/bookinfo

Sensors and Switches

Sensors and **switches**[20] are devices which detect some condition in the physical world and report that condition's status to the control system. A sensor or switch could be a button pressed on a control console by a human operator, a photo-electric beam crossed by a performer, a proximity switch that senses when a metal machine part is near, or a sensor measuring the air temperature. There are literally thousands of devices available for all kinds of applications, but let's take a look at a few kinds that you might commonly encounter in a show control system.

Operator Controls

This category means any control actuated by an operator. This can include anything from an industrial control actuated by a theme park employee to a "go" button on a lighting console. There are literally thousands of types of controls for just about any application, depending on the application, the contact arrangement required, etc. But my advice is not to scrimp in this area; a cheap button could stop your entire show.

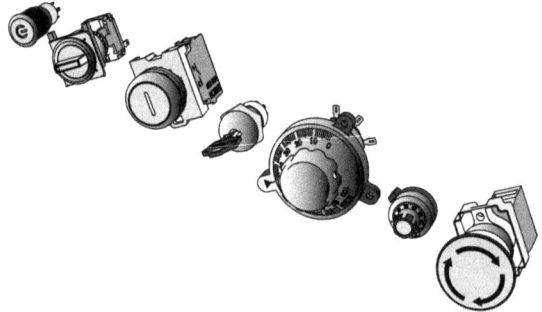

Wireless Remote Controls

Wireless (typically radio or infra-red) remote controls are another form of operator control that are useful in many types of show control systems. Handheld units for our market come from the security or industrial control industries, where actuation reliability is taken seriously, and can be useful where a client or performer wants to trigger an effect discreetly, control a private entrance, and so on or, of course, in any situation where wires are not desirable. Generally, there is a base unit that contains a receiver and the electrical interface, and then a transmitter. For reliability reasons, I always advise using wireless anything only when there is

---

20 I have not been able to find any good formal definitions describing the difference between sensor and switch. So, for the purposes of this book, a switch involves something mechanically operated, while a sensor perceives conditions in its environment in other ways. So a mechanical limit would be a "switch," but a temperature probe would be a "sensor."

no way to use a wired interface.

Limit Switches

A limit switch is typically actuated by a machine part that physically contacts the switch, indicating that a "limit" has been reached. Limit switches offer absolute positional control and come in a variety of designs, ranging from rotary-cam limit switches to simple mechanical switches mounted on a track or piece of machinery. Both are actuated when the a mechanical device hits a "limit," or desired target position.

Photo-Electric Sensors

Photo-electric sensors typically send an infra-red beam of light directly or indirectly (via a reflector) to a receiver, which then sends out a control signal when the beam is either interrupted or made, depending on the application. These sensors are useful in a wide variety of show control applications in which you want a performer or audience member to activate something by simply breaking an invisible beam.

"Light operate" sensors turn on their outputs when light is detected; "dark operate" sensors turn on when the light beam is broken. "Retro-reflective" sensors utilize a beam that bounces off a reflector, and are so named because the reflector gives back light at the same angle from which it arrived.

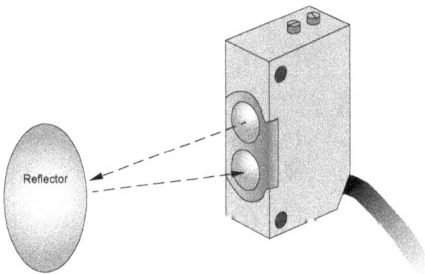

"Thru-beam" photo sensors are those that send a beam from a transmitter to an active receiver.

Motion Detectors

While most of the sensors we've discussed so far have come from the world of industrial control and factory automation, motion detectors typically have their roots in the world of security and alarm systems. These devices typically send an infra-red light, microwave radio wave, or ultrasonic audio signal into a space; if someone moves through that space, they disturb the field and the motion can be detected from the change in reflections.

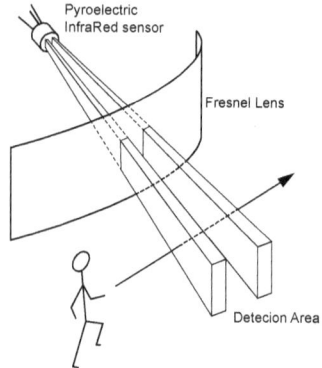

Systems are also available that monitor a video signal and look for visual change. Whatever the technology, these sensors can be useful for a variety of show control applications—from haunted houses to museums. However, you should keep in mind that these sensors are typically designed with reliable actuation—not necessarily quick response time—in mind.

Encoders

An **encoder** typically takes a physical position, and senses and "encodes" that position into some kind of data readable by a control system, typically in a digital format easily readable by a computer. These devices can be used to sense things like the position of a piece of scenery or a knob on a control console.

Other Kinds of Sensors

We've only scratched the surface here of what's available. Sensors exist for nearly any physical characteristic you can imagine: temperature, pressure, velocity, force, strain, position, flow, and so on. These systems might have a simple on/off or even output data over a network.

Direct Outputs

We've covered various ways to sense basic on/off information from the outside world and get it into a show control system; let's look now at ways to have our system directly control other on/off devices (keeping in mind, of course, that—like inputs—it's also possible to have a remote output device accessed via a network).

Relays and Contactors

While computers and controllers can typically produce small electrical currents for control, we often want to run larger loads. A simple electro-mechanical device that will do this is a **relay**, or a **contactor** (contactors are typically large relays handling larger power applications). Both work in the same way: A small control current is used to energize an electromagnet, which, in turn, physically pulls a set of contacts together, closing the circuit.

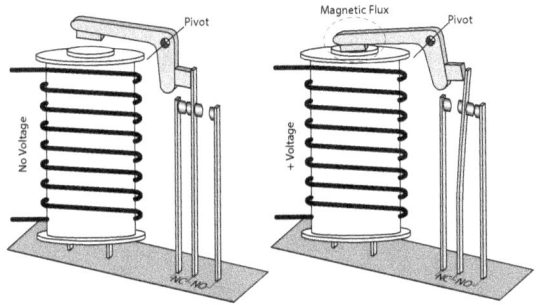

These contacts can be designed to handle very large amounts of current and, therefore, allow large, high-power devices to be operated from small control signals. Relays are mechanical devices, which means they will (eventually) wear out, but they are easy to interface and good for switching things that don't have to turn on and off super fast.

Transistors and Solid State Relays

Electromechanical devices like relays are solid and robust devices that have been around since the 1800s. However, as mechanical devices, they can wear out, arc under heavy loads, be noisy, and are not capable of being switched on and off quickly. For high speed, low cost, (relatively) lower current applications we have solid-state switches with no moving parts: transistors. These can take many forms in control systems, and there are thousands and thousands different kinds of devices available, and lots of resources to help specify them.

A Note On Electrical Isolation

When you connect electrical devices or systems—especially low voltage and high voltage systems—the potential is created for all sorts of nasty things to happen. What happens if high-voltage, high-power electricity (such as that found in a motor drive) connects to low-voltage electronics (such as that found in a computer)? Fire and smoke, typically! The easiest way to overcome these problems is to break the path for damaging current or, in other words, to provide electrical **isolation**. Isolation devices are usually inexpensive and expendable—if a significant fault occurs, the small, inexpensive isolation device may burn up, sacrificing itself to protect the whole system. For all these reasons, isolation is cheap insurance when interfacing outside devices to computers, which we do a lot of in show control. There are two basic ways types of isolation: optical and galvanic; when interfacing with higher voltage, real world equipment, it's always good practice to ensure that your system has isolation.

MISCELLANEOUS PROTOCOLS

There are a number of generalized networking protocols that might find applications in show control systems that are worth a mention here.

MQTT
MQTT is a lightweight messaging protocol often used in the Internet of Things (IoT) market.

Positional Tracking and Interchange Protocols
There are a variety of methods to communicate positional information in more or less standard ways, depending on the solution. These approaches enable applications like allowing a moving piece of scenery to be tracked by a projection mapping system, or a performer tracked by a laser projector. Approaches for this include PosiStageNet, the Real Time Tracking Protocol (RTTrP), NatNet, and ESTA's Object Transform Protocol (OTP).

Simple Network Management Protocol (SNMP)
The Simple Network Management Protocol (SNMP) was developed for—not surprisingly—network management. With the increase in network use in entertainment, SNMP has found a place in configuring entertainment devices as well as IT equipment (routers, switches, etc.) SNMP communications take place between a network-management system, or "manager," and a device element, or "agent," which is traditionally a router, hub, network printer, and so on.

Network Time Synchronization Protocols
There are two primary ways of synchronizing disparate devices over a network, at different levels of resolution: Network Time Protocol for general synchronization, and Precision Time Protocol for precise applications.

Network Time Protocol (NTP)
The Network Time Protocol (NTP) was developed in the mid-1980s to synchronize machines across a network (including the Internet). The protocol can synchronize free-running, local computer times to typically within a few milliseconds. This is very useful for time of day and even cue-level synchronization, but not really precise enough for audio/video clock synchronization (for that, see PTP below). Keep in mind that NTP is a synchronization tool, not a clock itself, so it's not a replacement for a "show clock" implemented in something like SMPTE or

MIDI Time Code.

Precision Time Protocol (PTP)

The Precision Time Protocol (PTP) standard was released in the early 2000s, can synchronize clocks over a local network to sub μs (microsecond) precision, which is accurate enough for audio/video clock synchronization. For this reason, PTP is used as a synchronization tool for widely used, local audio streaming approaches.

MOVING ON

There are a huge variety of devices that can be connected to show control systems, and just as many ways to connect them, but you can almost always find a way to interface with something, especially if it has an Ethernet jack. Let's move onto my process of show control system design.

Chapter 3

A SHOW CONTROL DESIGN PROCESS

SYSTEM DESIGN PRINCIPLES

Before moving on to the details of a show control system design process, let's take a "big picture" view of general design considerations for show systems. Everyone designs things differently; what follows is a list of the general principles and objectives I use when designing any kind of system:

1. Ensure safety.
2. The show must go on.
3. Simpler is always better.
4. Strive for elegance.
5. Complexity is inevitable; convolution is not.
6. Make it scalable, and leave room for unanticipated changes.
7. Ensure security.

Of course, these are highly generalized principles and, as we know, there are always exceptions to any rule (except Principle 1: Ensure safety).

1. Ensure Safety

"Safety is no accident" is the cliché, but this one is true. Safety can only exist if it's not an afterthought—it must be considered for every situation and action, from the top of a system, process, or design to the bottom, from the beginning to the end. Safety must always override any other consideration: If you can't afford the resources (time, money, and so on) to do the project safely, you can't afford to do it at all.

There are a variety of safety standards that cover many issues related to show production, and you should always check with the **authority having jurisdiction** (AHJ) to see which apply to your situation. Beyond that, there are a few principles that I follow in my designs. Use these if you agree but, of course, I make no claim that following these principles will make anything safe! Safety is up to you—the system designer.

Fail-Safe Design

Fail-safe design means that a system is designed to fail in a safe way. For example, relays or contactors can be wired so that in case of a control system power loss, the connected device is switched to a safe state (typically de-energized). In a similar way, some mechanical brakes are controlled using air pressure, but the air pressure actually *releases* the brakes, rather than applying them. That way, if a brake hose becomes detached or develops a leak, the system will fail into a safe condition, applying the brakes.

Single Failure-Proof Safety

Another good design principle is **single failure-proof design**, where a system is designed so that if any single (anticipated) failure occurs, the system will not fail in a way that creates more danger. The classic example of this principle in the entertainment industry is the lighting safety cable.[1] If the lighting fixture falls for any reason, the safety cable keeps it from hurting anyone. Of course, the light may end up pointing in the wrong place, and you may need to get out a ladder to get the light down, but the goal here is safety, not performance. Single failure-proof design can also help to increase reliability through redundancy.

Emergency Stop Systems

Emergency stop, or **E-Stop** systems, are critical for any dangerous control system that could injure someone. An E-Stop generally works by disconnecting a drive device's power at very low levels, or somehow safely disabling the drive capability. When in emergency stop mode, the system is unable to operate, no matter what the control system instructs it to do. Related concepts include a "safety relay" or a hardware mode of some devices called "safe torque off". E-stop systems and related systems are generally constructed with simple or safety-rated components and kept independent of the main computerized or electronic control elements. Once put into an E-stop condition, systems typically latch into the stopped state and require some sort of reset before enabling power again. Control systems should be designed so that they do not resume movement immediately upon exit of the emergency stop state. This is a complex topic mostly outside our scope here, but these concepts are important to consider in show control system design.

Humans in the Loop

In my opinion, humans should *always* be in the control loop of any entertainment effect or system that could hurt someone. In factory automation and robotics, they

---

1 For those not familiar, the safety cable is a strong, rated wire rope cable looped around the hanging position (typically a pipe) and through the "yoke" of the unit, or onto a specially made hook built into the unit.

build big fences around dangerous robots and add interlocks so that if someone opens the fence gate, the machine stops. This approach is useful in some entertainment applications (e.g., under a stage), but there are many effects, particularly on a stage, that can't be fenced in without ruining the show. Computerized control systems can apply tremendous processing power to get highly precise effect timing and cuing; but any dangerous effect must never be allowed to operate without some positive action from a human somewhere—on or offstage. **Enabling** systems are often used to incorporate this human intelligence into control systems.

Enabling systems allow humans to "authorize" or "enable" dangerous effects. There are many ways to implement these ideas, but let's look at a few examples from the theme park world, where extremely dangerous effects are fired in close proximity to people dozens of times a day. With our first approach (see diagram below), a computer could continuously monitor a number of safety sensors and, if all the sensed conditions are safe at the appropriate time in a show, open a timing "window" that enables the execution of a dangerous effect.

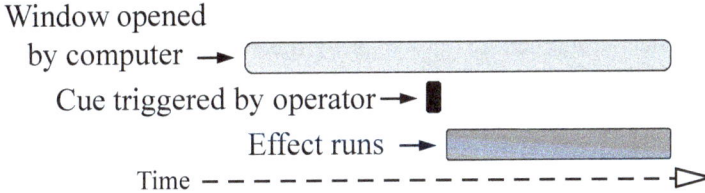

When this window opens (typically for a limited amount of time), the effect is enabled and this condition is indicated to the operator in some way (e.g., a button is lit). The operator can then fire the effect (push the button) at the right moment in the show after confirming that conditions are safe. The actual job of the operator in this case is to *NOT* fire the effect if things are not safe for the performer.

Another control structure would be for the operator to enable the computer system (typically by pressing a button on a console) when everything is safe and ready; the computer then can fire the effect at a specific time in coordination with other systems.

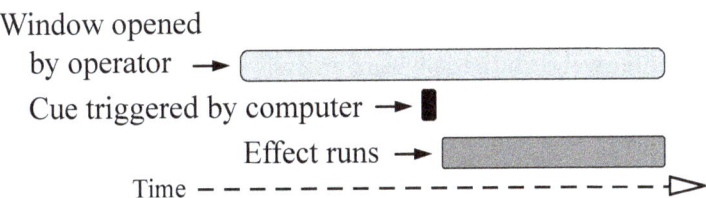

Still another possibility is that the computer could open a timing window, an operator would enable the effect, and the *performer* would actually fire it by stepping on a switch or some other means.

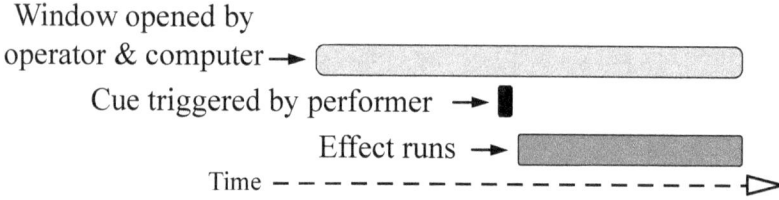

Operator Issues

On long-running shows, backstage operations can become routine, so system designers should take operator boredom or complacency into consideration, especially for anything with a safety implication. Well-designed enabling systems, for example, typically look for some sort of *transition* (button pressed, button released, or button pressed and released) from the operator's control, not just a steady state. This keeps a bored operator from placing a rock on the enable button in order to take a break. If the enable is not seen by the system at the appropriate time, the effect can be programmed to fail in a safe manner or go into some sort of "graceful abort" mode.

There are many common-sense techniques to counteract operator fatigue, illness or complacency. For example, you could place part of the operator's control or indicating system *inside* the show or (safely) near the dangerous effect, forcing them to look that way or otherwise be paying attention. Alternatively, this idea can be implemented with accelerometers and gyroscopic sensors that detect when a hand-held device is held upright; if the device is dropped, held still too long (cheating the system) or tilted, it means the person is no longer paying attention. A related fail-safe idea "driver's safety device", "operator presence control", or, historically, the "dead man's switch", is used on things like subway trains. In order to make the train move, the operator has to press down on a control; if they are taken ill or leave their post, the train stops.

Hazard Analysis

Hazard Analysis is the formalization of a safety technique that system designers often use intuitively. The basic hazard analysis process anticipates possible failures that could cause unsafe conditions, and then evaluates the severity of each possible failure along with the likelihood of occurrence. There are several standardized methods for this process, depending on the application.

Peer Review for Safety
No one can see every eventuality, anyone can make mistakes, and when you're deep into a design it can get hard to see the details after a while. An excellent approach to counteract some of this is to have a qualified peer review your work. Ideally, of course, this should start when the project is still in the design phase, when things can be easily changed.

2. The Show Must Go On
Our show is the product that we sell; that's why this axiom is second nature to those of us in the entertainment industry. Show systems must be designed to be flexible and quickly and easily understood, diagnosed, repaired, or, even bypassed. An operator must be able to run the system manually if necessary, and it's critical that backup systems are in place and are easy for the operator to use, even when under enormous pressure.

Redundancy
One of the most straightforward ways to ensure that the show goes on is to make sure there is redundancy in your system. Compared to the cost of a lost show, computers and related devices are relatively inexpensive, so if you're running a sound effects system, it's quite easy and often economical to have a backup machine. If the main machine dies in the middle of the show, you can simply switch over to the backup. How smoothly this transition goes typically depends on how the system is designed, and in networked systems, how carefully the implications of switching over have been analyzed. If you can execute the transition seamlessly, without losing your place in the show, this redundant approach is typically called a "hot" backup system. How much redundancy is ideal or practical depends on the situation and the budget.

Designed Boundaries
Part of being in the show technology business is dealing with pressure. There's nothing like having to troubleshoot something while thousands of people are waiting for the show to start, or money is burning in rehearsal. One technique to facilitate troubleshooting is to have clear boundaries between systems. For example, if you have a video playback system connected to a show control system through a network, that network interface is now a boundary. Testing can be done at that boundary, or sub-systems can be quickly replaced or interchanged.

Ensuring Maximum Computer Reliability
While of course reliability is a concern with every kind of device in a show control system, we are often dependent on general-purpose computers and networks, so we should take some special precautions. I go over this further from the network

side in the *Introduction to Show Networking* book[2].

Statements about which operating system "never crashes" are pointless and inevitably lead to ridiculous, drawn out flame wars; every system is capable of crashing (and if it's not the operating system, a hardware failure is always possible), so it's best to plan your system design with that in mind. Following are some general good-practice principles that I apply to any computer running a system on a show.

Keep Your Show Network Off the Internet Unless You Have No Choice
Connecting your show network to the Internet, in addition to opening your show up to a world of security issues (see System Design Principle "7. Ensure Security" on page 57), gives all kinds of software processes on your machine the opportunity to update themselves, register, report on your usage, and so on, which may disrupt critical show events. If your show network is physically disconnected from the Internet, these processes are far less likely to pop up in the middle of your show. If you do need to connect your show network to the Internet to update software or something similar, then try to do it only when needed and then disconnect your network again. If you have to connect your network to the Internet either because of the needs of the show or for remote maintenance or programming purposes, one approach that is commonly used is to put a machine on the local show network and equip it with two physical network connections—one local and one remotely accessible. This makes a single point of entry to the show network that can be tightly controlled and managed.

Install Only What You Need, Remove What You Don't Need
In the context of a show, even a powerful computer is relatively cheap, so you should dedicate your machines for show purposes only—they should contain nothing but the software you absolutely need for the show. Check with your software vendor for operating system tweaks and optimizations. For similar reasons, do not install games or other "fun" software on the show machine—bring your own laptop or phone for that purpose.

Manage Automatic Virus Checking
In the 1990s, if your network was off the Internet, and your operators were following good practices, there should be no way for a virus to "infect" your machine in the first place, and virus checkers placed a processing load on the machines so they were often discouraged. But when a system must be connected to the Internet, there is some vulnerability, and you have to make a judgment about your risks and decide what you want to do. However, virus checkers often operate in an

---

2 More info here: https://www.controlgeek.net/bookinfo

automatic mode, and you really don't want them deciding that it's a good time to thoroughly scan your disk while in the middle of the most demanding cue of the show. So it's best to carefully manage the settings on your virus checking software, and follow good practices about what kind of removable media and devices you allow to be connected to your show computers.

Store Your Show Files Locally

It's great to have a local network (or cloud based storage solution, if managed carefully) available to backup your show files or to load media, but in my experience, it's generally a bad idea to actually pull your show files from a drive over a network, since that opens you up to potential failures if the network has a problem during the show. Use a local drive instead.

Back Up Your Data

I am fanatical about backing up. It's always surprising to me how even experienced people can be cavalier about backup, when your data is your work! If you lose a day's changes to your show, you have lost a day's work. I typically keep backups on multiple drives in different locations, and I save a new version (or backup version depending on your software) of the show file itself each day (with a date in the file name like YYYY-MM-DD), or maybe even multiple versions within a single day. This way, if I totally screw up my file (or—worst of all—overwrite a good file with bad or corrupt it), I can always go back and get yesterday's file, or even last hour's file. Keep an up to date copy of your data offsite or in the cloud. Storage is cheap and there is no excuse for not backing up regularly and often. Keep in mind that data doesn't have to be lost by some mysterious cause; it can just as likely be lost when someone spills coffee on the machine, frying it.

If It Ain't Broke, Don't Fix It

Most general computer programmers write code, debug it, and then release it. Barring a catastrophic business failure, if they break something in this process, it's unfortunate, but it's not likely going to disappoint thousands of people who came to see a show. So, given the ephemeral nature of a show, my advice is to follow the old adage of "If it ain't broke, don't fix it", and don't mess with a working system, at least until you have time to make the changes and test it out in every way you can imagine. That improvement or little tweak may seem harmless, but there could be unintended consequences; wait until after this show run to update that software or replug that connector.

Document, Label, Update!

Much of our critical system design is implemented in networks and abstract tables of numbers that take hours to create, so it's critical to document every setting or configuration you make for a show, label everything, and update all of this regularly. And it's so easy to with camera phones, online spreadsheets and documents, and label technologies that you really have no excuse. And even if you *think* you can remember everything and don't need to document it, keep in mind that your memory may not be as good as you think, and in our show-must-go-on industry, someone else may have to step in on short notice to fix your system.

3. Simpler Is Always Better

I have never seen a situation where a more complicated solution is better than a simpler system that achieves the same result. Simpler systems are easier to design, install, program, and troubleshoot (and all that generally adds up to less expensive). No one will ever be impressed by a complex system that no one can troubleshoot, no matter how many blinking lights it has. If it's possible to cut something from the system without compromising performance or decreasing flexibility, get rid of it.

4. Strive for Elegance

An elegant solution uses a minimum of resources to accomplish a task, often in a surprisingly simple way.

5. Complexity Is Inevitable, Convolution Is Not

Big systems are inevitably complicated, but there is *never* a reason to have a convoluted system—one more complicated than necessary. Convoluted systems are generally the result of poor design, planning, implementation, or documentation, all of which lead to an unsightly mess that no one can ever figure out. Everyone I've ever known who has created convoluted systems as a means of "job security" has been eventually fired when they were not able to fix the system—sooner or later even they couldn't decipher their own convolution. To paraphrase the old quote[3]: Everything should be made as simple as possible, but no simpler.

6. Make It Scalable, Leave Room for Unanticipated Changes

The cost of a piece of cable is extremely low relative to the costs of designing a system that needs the cable and the labor involved in running and terminating it. Always run spare cables, order more than you think you need, and buy spare parts. Do this even if you cannot imagine any possible way the system could need expansion—someone will soon figure that out for you. I've never heard anyone complain that spare capacity was in place, but I have heard plenty of grumbling about not

---

3 Often attributed to Einstein, but this isn't established with any certainty.

having room for expansion. While it's often difficult to convince bean-counters, it's much cheaper to put in room for expansion now rather than later.

7. Ensure Security

With so much riding on our systems, we have to make sure that they are secure. Security applies to all types of devices in a control system, but especially, of course, to computers and networks; I have more on this from the network side in the *Introduction to Show Networking* book[4].

Following is a basic list that I use for any computer-based, networked control system. Some of these overlap with what we discussed in System Design Principle "2. The Show Must Go On" on page 53, but we're looking here from the security perspective.

Show Networks and the Internet

It's generally a lot tougher for an attacker to get into your network if it's not connected to the Internet. If you need to connect your network to the Internet for a software update, be sure to disconnect it again before show time. If you must keep your network connected to the Internet, make sure you have a clean, limited connection point, learn everything you can about firewalls, and use one. Also consult a talented IT security expert.

Use Security Features on Wireless Networks

There are so many portable wireless network devices out there now, it's impossible to know who might try to access your network just for fun. Be sure to encrypt or hide any wireless networks you might have.

Shut Down Wireless Networks When Not Needed

It's great to use a wireless tablet to move around the venue and adjust things during rehearsals. However, if you are not using wireless during the show, disconnect or power down your wireless access points.

Use Passwords on Critical Machines and Systems

This simple, but often overlooked, step will stop people from casually trying out a control screen they happen to encounter.

Control Physical Access to Critical Infrastructure

Simply locking your critical equipment up can make it very difficult for outsiders to get access to your network (if it is to be an inside job, they will likely have access anyway).

---

4   More info here: https://www.controlgeek.net/bookinfo

Run Only Things Needed for the Show on Any Show Machine

That fun website your operator might like to visit may also contain all kinds of viruses, spyware, and so on. Tell the operator to bring in a laptop or their phone if they want to surf the net or play video games.

SHOW CONTROL SYSTEM DESIGN PROCESS

So, considering everything we've discussed so far, let's move onto a show control system design process. There is no official or standardized way to design a show control system; additionally, there are many ways to achieve any desired functionality. But I think we can approach the process systematically, and here's the questions I ask when first approaching a show control application, based on the concepts we discussed in Chapter 1, "Introduction" on page 1, along with some additional details:

- What are the safety considerations?
- What type of show is it?
- What kind of triggers and synchronization should be used?
- What devices or systems are to be connected/controlled?
- What system architecture works for this application?
- What is the control information source?
- What type of user interface is required?
- What kind of show control approach is needed?

Answering those questions will sort out most of the design of the system; after that it's mostly programming and debugging. Note too that I'm presenting these here in an order, but depending on the circumstances it might make sense to do the steps in a different order, or even combine several steps together. Let's go through each of these questions in some detail; afterwards, we will apply this process to some case studies.

What are the safety considerations?

Safety, of course, must be the utmost consideration in any situation, as we discussed in my System Design Principles, "1. Ensure Safety" on page 49. Many entertainment technology systems are basically safe for audience and crew in all but extreme circumstances: lighting, sound, video, and so forth.[5] Other systems, such as scenic automation, pyro, and flame systems, are very dangerous and must be approached with extreme caution—you should get expert advice before integrating with them. But with proper safety precautions and procedures in place,

---

[5] Of course, you can electrocute someone, damage their hearing, etc. but you get the idea.

these systems can and often are integrated using show control to provide highly precise timing accuracy, enhanced safety, or other functionality not possible with human operators.

What type of show is it?
As we discussed in "Show Types" on page 5, our basic types are linear and non-linear.

What kind of triggers and synchronization should be used?
From "Triggers and Synchronization" on page 6, the three basic types are event-based, time-based, or hybrid event/time based.

What devices or systems are to be connected/controlled?
Since show control, by definition, connects systems, it's not possible to complete the system design until you know about all the systems or devices that are being linked together. We discussed a range of types of gear in "Entertainment Control Disciplines and Devices" on page 16. Each controllable device or system will accept one or more control protocols/standards over one or more connection methods. Selecting the right control approach for each device (or, selecting the right device based on how it can be controlled) takes some thought about the design. I often start this process by making a spreadsheet showing each device and system, including what control ports each device has (leave plenty of room in the spreadsheet for network information) and what protocols can be sent over that control port.

What system architecture works for this application?
As we discussed "Show Control System Architectures" on page 13, there are a number of architectures and structures that can be implemented. Picking the best one for a particular system means considering everything from interdepartmental politics to budget, and from the available equipment to the specifics of the venue.

What is the control information source?
This may sound complex, but what we're looking for is simple: We need a way to get information about the show into the control systems. Is a human operating the system? Are the audience members themselves triggering the cues? Is the performer triggering effects? Are we tracking a moving platform? Is the show run from a time-of-day clock? Does the show start at sunset and run through sunrise?

Determining this and how you can get this information into some machine-readable form will be a big factor in deciding which system or technology is best for your application. We discussed a number of ways to get data and information into your system in Chapter 2, "Connecting Devices and Systems" on page 21.

What type of user interface is required?
If the system is run by skilled technicians comfortable with cues, protocols, and timing issues, they may want as much user interface, feedback, and status information as they can get. In other applications that are simpler or more routine, there may be only unskilled operators who might be confused by too much detailed feedback and status information. For example, in a retail environment, someone who knows a lot about shoes but nothing of show technology may be starting and/or selecting the shows, and much of the editing functions may need to be completely locked down, and maintenance features controlled via hidden screens. Each of these applications needs a different kind of user interface, and you need to determine this in figuring out your approach. We discuss some operator issues in "Operator Issues" on page 52.

What kind of show control approach is needed?
Now with the answers to all those questions in mind, we can develop our approach for the system. There are many ways to coordinate systems; you can do show control without a "show controller", or for more sophisticated applications, you might need more capability. Let's dig a bit deeper into those ideas.

Show Control Without A "Show Controller"
This type of show control is often the simplest and most straightforward to implement. You might run time code from one system to another (or a few) with a cable, having all the subsystems chase the leader in a primary/secondary structure. Additionally, you might simply connect two devices on a network, and have them exchange messages in a peer-to-peer fashion. Sophisticated show control can be implemented this way.

When You Need a Show Controller
There are many applications where simple primary-secondary or simple peer-to-peer interconnection is not enough: perhaps your show needs some special conditional logic; perhaps there are inter-departmental political reasons why it makes more sense to use an intermediary system to react to changes as they happen; or perhaps one of the systems doesn't allow easy, time based editing, doesn't have a time of day clock, and so forth. In these cases, you would need a "show controller".

Of course a show controller could take several forms; it could be software running on a general-purpose computer, or it could be a dedicated hardware show control unit (which typically is, of course, software running on a dedicated computer). For me, a full-featured "show controller" has a few key characteristics:

- The system must be able to communicate bidirectionally using a variety of protocols, ideally over multiple networks.
- The system should be able to output and receive contact closures and other similar I/O methods (if this functionality isn't built in, it can be implemented remotely over a network).
- The system must be able to work in any **interrupt-driven** way, meaning that it can act on any information coming from anywhere at any time and doesn't have to wait for a poll or other similar command.
- The system's program must be able to be edited while the show is running (this is a big one).
- The system must be able to precisely manage multiple asynchronous (independent) timelines, and sync to multiple timing sources, external and internal.
- The system must be capable of stand-alone, automated operation (e.g., working off a time of day clock, etc.) without any human intervention (for fixed installations).
- The system should be able to present a customizable user interface.

A Note on Media Control Systems

For show applications, be wary of the "media control" touch-screen based systems used in many corporate boardrooms, restaurants, and home theaters. For simple, operator-controlled applications, these systems are powerful and cost effective, and for that reason, specifiers are often tempted to use them for show applications. However, you should do your research carefully; these media control systems (while sometimes used in shows for user interface purposes), rarely offer the timing accuracy/repeatability or the flexibility inherent in true show control systems. In addition, these systems are often programmed using a compiled, general-purpose programming language interface, meaning that programmers must create from scratch much of the functionality of real show controllers. More importantly, media control systems typically do not offer the programming flexibility and speed needed to survive the "edit on the fly" live show environment. [6]

---

6 A programmer of media control systems asked me, "How long does it take to program XYZ show control system?" I asked, "How much time do you have?" They were baffled. "If you have a day, it's a day", I said. "If you have a month, it's a month." The programmer usually based their programming estimates for on the number of buttons needed to create the touch screen. Eventually, after much explaining, they got it—this is a show, not a boardroom, and the programmer (wisely) declined to bid the show project. This has come to be known as "Huntington's Law".

A Note on Physical Computing Systems

There are a number of small micro controller-based embedded control systems that offer amazing power and flexibility at a very low price. These systems can be very useful for prototyping and low-level subsystem interfacing to a show controller; however, if you are basing your system on one of these devices, you are now the low-level hardware engineer and the programmer, responsible for all aspects of the interface, top to bottom. Keep that in mind too that many of these systems were not designed for the harsh electrical and physical environments found in many show applications, and may not be approve-able by electrical inspectors and other Authorities Having Jurisdiction. Additionally, few of these systems can be have their code changed while operating.

A NOTE ON BUDGET AND TIME

I left budget off the question list in my design process because I firmly believe the most effective design process is to first figure out what is truly needed by the story you are trying to tell, and see what that costs. If you can't afford to do it right, then cut something. Extensive "value engineering" to meet an inaccurate budget estimate often results in a system that appears to work on paper, but fails to meet the requirements of the show. If a client is insisting that you compromise your system to the point of ineffectiveness in order to meet a budget goal, you probably won't regret (in the long run, anyway) walking away from the project.

In addition, one major item to consider in your budget, of course, is time. Show control offers enormous benefits, such as increased precision and repeatability, and even freedom and flexibility for the performers. However, those enormous benefits come at the enormous cost of a large amount of programming, testing, and tech time. You also need to leave a lot of time for testing and debugging; we are *programming* after all. Be sure to include the impact of that time in your budget.

DO YOU REALLY NEED SHOW CONTROL?

As a final wrap up to your design process, I would recommend considering one additional question: Do you really need show control for this show? As I've said, I believe that the simplest solution that gives the desired results is best (System Design Principle "3. Simpler Is Always Better" on page 56). Is show control adding unnecessary complexity, or worse, convolution? (System Design Principle "5. Complexity Is Inevitable, Convolution Is Not" on page 56). If it is, you should rethink whether or not there is a simpler way to get the job done.

Done right, show control is capable of offering a level of sophistication in your show not otherwise possible. But if you don't have the necessary time and other resources available on a particular show, don't even attempt show control. Try some experiments in a no-pressure situation first to see if it will work for you.

MOVING ON

Now, to bring it all together, let's take a look at some practical examples of show control systems, applying this design process.

Chapter 4

EXAMPLE SHOW CONTROL SYSTEMS

In this final part of the book, we'll bring together everything we've covered and look at a few practical (albeit hypothetical) show control problems and solutions. Please note that I chose the examples, systems, and approaches here to illustrate specific points, so please take these systems only as possible solutions to the design challenges presented, especially since one of the most fascinating (and simultaneously daunting) aspects of show control is that there are many ways to achieve any particular design goal. For each example, we will go through, step by step, my design process detailed in "Show Control System Design Process" on page 58, keeping in mind my general principles from "System Design Principles" on page 49.

ENTERING THE ARENA

"Lets get ready to roll!" booms the announcer over the public address system. "Please welcome," the announcer continues, "number seven, Logan Landry!" Rolling into the arena, we see Logan's scoring stats on the video screen, all set to a thumping beat. "Up next," the announcer continues, "Number 66, Timber Hollis!", and as Timber rides onto the floor, lights flash as we see on video that they are the top scorer in the league. Meso, the team's mascot, goofs around with each player as they come into the arena.

The Mission
The Fighting Tornadoes unicycle basketball team at Vorticity high school wants to spice up their team entrance for their games, and they approach A/V club president Kit Landry to see what's possible.

What are the safety considerations?
While the unicycle riders of course take on a bit of injury risk, the lighting, sound, and video systems that need to be connected here don't really pose any intrinsic safety risks.

What type of show is it?
Since the team members always roll in in the same order, the show is linear.

What kind of triggers and synchronization should be used?
To keep things simple and make it easy to program a series of fast-paced lighting cues precisely timed to the audio and video, Kit designs the system to be be time-based.

What devices or systems are to be connected/controlled?
Kit is comfortable working in a software-based audio/video playback and show control system which can send or chase SMPTE or MIDI Time Code, or communicate via a variety of protocols over a network. While this software-based system can also control lighting equipment, in order to quickly create some flashy moving light cues, a lighting console is "borrowed" from the high school's auditorium. This console also can receive and chase SMPTE or MIDI time code, and have any of its parameters or cues controlled via OSC.

What system architecture works for this application?
In this case, since the show is time based, and the AV playback system has flexible time code control options, Kit decides to use a simple primary-secondary approach, with the AV playback system driving the lighting console.

What is the control information source?
The control information for this show is the AV playback software itself, which Kit will run for the show.

What type of user interface is required?
The AV playback software's built in user interface contains quite a bit of feedback and status information which makes monitoring the show easy, and since the AV club members will be running the show, they want as much flexibility as possible. Kit plans the show to start with a single Go button press.

What kind of show control approach is needed?

Kit and the club meet and come up with a system design, and get everything tested out. With this relatively simple system, all that is needed to implement show control is a single XLR cable between the two systems carrying SMPTE LTC.

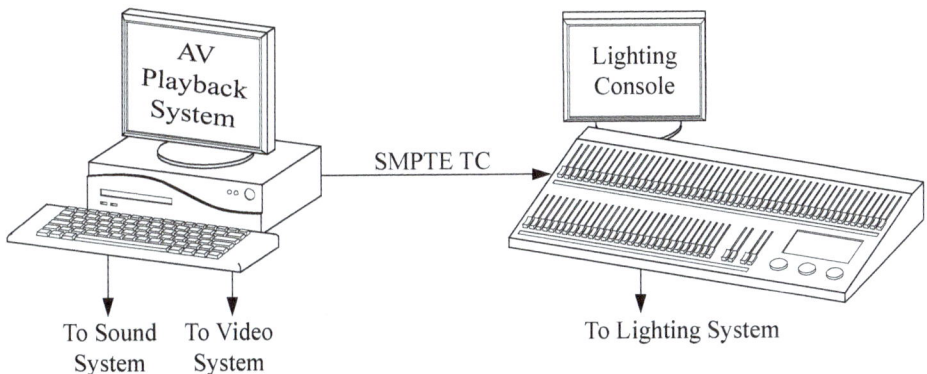

Kit and the team get everything configured and then lay out the timeline as follows:

Trigger	Event
Manual Go	10 seconds pre-roll from 00:59:50:00
01:00:00:00	Start of show, music and video starts
01:00:00:00	First light cue, music starts
...	Series of lighting cues triggered locally on lighting console
01:03:28:12	End of Show, static light look
Manual	Lights fade out

While everything tested before the audience comes in, the ten-second pre roll gives the operators confidence that the time code is running properly. Once the show starts, the audio and video runs, while the lighting console triggers pre-programmed cues at specific times in tight sync with the audio and video.

The show runs great in rehearsals, and with a single Go button press on the A/V software a spectacular lighting and video show is run in tight synchronization with the sound track.

Refinements

Since the AV Club worked on this well in advance, they get the show dialed in and have time to make a few enhancements. They add in a network switch and connect the two systems together with Ethernet:

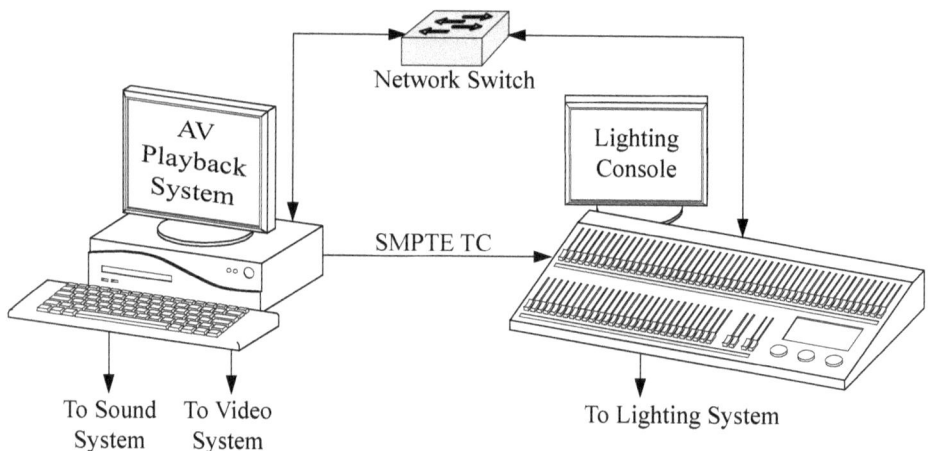

This additional functionality is added because in order to set up the show, the lighting console needed to be in a first pre-show cue 99. With this new structure, Kit adds in via OSC a /cue/99/go, targeted at the IP address of the lighting console. While this does add some unneeded complexity to the system, it adds flexibility and the lighting console can always be run manually if necessary.

Thinking it through and rehearsing with the teams' announcer, Kit realizes that ten second pre-roll is too long. So, in order to make the "Go" a bit tighter, the AV team changes the sequence to pre-roll and then hold one second before show start. The remaining one second of TC after the "go" allows the lighting console time to reliably sync up. The actual pre-roll time is extended just to give a more solid test:

Trigger	Event
Manual	29 seconds pre-roll from 00:59:30:00
00:59:59:00	Automatic Pre-Show Hold
Manual Go	Start of show, music and video starts
01:00:00:00	First light cue, music starts
...	Series of lighting cues triggered locally on lighting console
01:03:28:12	End of Show, static light look
Manual OSC	Lights fade out

In addition, Kit writes a pre-show check cue list that sends video to each output, audio to each speaker, and brings up each lighting fixture in the system to test it out. To fire the lighting console, a series of OSC cues are written into the pre-show list cue list.

The show is so well received that Kit and the team are hired by the local lighting shop to program shows.

A THEATRICAL THUNDERSTORM

A huge flash of lightning illuminates the stage and, two seconds later, "Kaboom!" the audience's seats shake as a huge thunderclap rolls through the venue. Ephraim Cubit, his three sons, and the girl next door are all huddled under the kitchen table as the storm continues. From the ever-shortening time between the lightning flashes and thunderclaps, we can tell that the storm is fast approaching. "I sure wish I had cut down that big, dead oak tree next to the house!," shouts Ephraim. As the word "house" leaves his lips, there is a blinding flash and simultaneous thunderclap. Seconds later, an oak branch smashes through the window upstage. As the sound of the giant thunderclap decays, Ephraim suffers one final indignity: His lights go out, leaving the stage in darkness.

The Mission

This musical version of *Desire Under the Oaks* is being produced at a regional theatre. The director will "spare no expense" to ensure that this sequence comes off perfectly night after night, although they have also insisted that the sequence not be "canned" (pre-programmed). We must keep in mind that "sparing no expense" in the regional theatre world is equivalent to sparing every expense at a theme park; solutions will be selected with an eye on keeping the cost as low as possible while still achieving the director's goal.

The show will run for four weeks. Lighting, sound, and props run crews are already on contract to the theatre, and most systems are already "in house." A show control system will be used only to ensure accurate timing for the tree branch effect; the rest of the show will be run manually by the operators on cues from the stage manager.

What are the safety considerations?

The scenic tree branch effect could be dangerous, and care should be taken to ensure that the effect cannot be triggered accidentally.

What type of show is it?
Because it's a theatrical production performed from a script that doesn't change from show to show, it's linear.

What kind of triggers and synchronization should be used?
Given the fact that the director doesn't want anything "canned" and that there are live actors involved, the system should be mostly event-based. However, the delays between lightning flashes and thunderclaps add a time-based element.

What devices or systems are to be connected/controlled?
Let's go through each one in detail.

Lighting
For the production, the electrician has rented a very bright strobe and, of course, the theatre already has a lighting console that controls all the fixtures and outputs DMX.

The old lighting console owned by the theater is capable of firing cues from MIDI Show Control commands. In addition to receiving MSC commands, the lighting controller also generates "musical" MIDI messages when its front-panel buttons are pressed. This could be useful for our thunderstorm, since MIDI Note On/Off messages are generated when the fader bump buttons are pressed and released.

The following MSC Lighting cues have been programmed into the board:

Cue	Effect	MIDI Command
218	Lightning strike 1	MSC Cue 218 Go
219	Lightning strike 2	MSC Cue 219 Go
220	Lightning strike 3	MSC Cue 220 Go
220.5	Big lightning strike	MSC Cue 220.5 Go
221	All lights out	MSC Cue 221 Go

The lightning strike "looks" from cues 218 to 220.5 have also been loaded into faders 1–4; when the fader bump buttons are pressed on this console, the appropriate lightning effect is triggered and a corresponding MIDI Note is sent out via the console MIDI port. The faders are programmed so that the lightning flashes last as long as the operator holds down the bump button.

Sound

In addition to a traditional musical-reinforcement sound system, the theatre has a MIDI-controllable sound playback system, which can be triggered either by standard MIDI Show Control commands or by MIDI Note commands assigned to each cue. The production sound engineer has programmed the following cues and set them to respond to both MSC and standard musical MIDI message:

Cue	Effect	MIDI Command
101	Thunderclap 1	MSC Cue 101/Note C2
102	Thunderclap 2	MSC Cue 102/Note D2
103	Thunderclap 3	MSC Cue 103/Note E2
110	Big Thunderclap	MSC Cue 110/Note F2
111	Tree Falling	MSC Cue 111/Note G2
121	Glass Breaking	MSC Cue 121/Note A3

In addition to the thunderclaps, the sound designer added two cues. As the branch effect falls, it doesn't make realistic tree crashing sounds, so its fall must be augmented with a "tree falling" sound effect. For safety reasons, stage glass is used in the window through which the branch crashes, and since this fake glass doesn't make much sound when it breaks, a "glass breaking" sound effect is also necessary.

The audio outputs of the sound FX system are connected to inputs on the main audio mixer, so the operator must remember to bring up those faders to the prescribed levels for the cue sequence. Since this is a live show, the sound mix position is in the back of the orchestra seats.

Props

The theater's technical director has devised a simple but effective way of electrically releasing the tree branch. The base of a scenic branch is mounted on a pivot and weighted, so that, without restraint, it will fall through the window. An electrical solenoid is mounted so that, without power, the branch is locked in the "out" position; when power is applied to the solenoid, the solenoid is activated and the branch falls (this is a kind of fail safe design—a power failure will not cause the tree to fall). Through experimentation, the TD has determined that the solenoid must be energized for at least one second to give the branch time to fall past the solenoid's shaft. In addition, a manual safety release pin has been installed parallel to the solenoid shaft, so that unless the pin is manually released, the tree cannot fall, and the area where the branch falls is blocked off. Once the actors are safely onstage and under the table, the pin is manually released by the prop crew, "arming" the effect. The solenoid is controlled by the light-board operator, using a push-button station positioned next to the light board.

What system architecture works for this application?
Since it's a simple system, the design team decides on a peer-to-peer approach.

What is the control information source?
The stage manager is in charge of the run of the show, and will call all the cues, including those in the thunderstorm sequence.

What type of user interface is required?
Skilled operators are present, and they will want as much user interface and feedback as possible. However, it's a relatively straightforward show, so it's likely that we can accomplish the goals with off-the-shelf systems, and it's not likely that custom user interfaces will be required.

What kind of show control approach is needed?
This is a relatively straightforward project; we can likely implement the systems without a separate, dedicated show control system. The stage manager precisely scripts out the whole sequence, working closely with the lighting and sound designers:

Actions	*Trigger*	*Control Messages*
First lightning strike	Stage manager	Light Cue 218
First thunderclap	Two second delay	Sound Cue 101
Verify safety of tree	Stage manager	Verbal Confirmation
Second lighting strike	Stage manager	Light Cue 219
Second thunderclap	One second delay	Sound Cue 102
Third lightning strike	Stage manager	Light Cue 220
Third thunderclap	1/2 second delay	Sound Cue 103
Big thunderclap/ Lightning strike	End of "house" line	Light Cue 220.5 Sound Cue 110
Tree falling/sound FX	Stage manager	Solenoid Go Sound Cue 111
Glass breaking	Tree through window	Sound Cue 121
Lights out	Stage manager	Lights 221

Approach 1

The design team wisely decides to test the entire sequence well in advance of technical rehearsals, using the equipment already on hand. The light board will be used as a MIDI control source, triggering the sound FX system; the tree will be triggered manually by the electrician using the push-button controller on command from the stage manager. All system interconnection is done via MIDI, as shown in this block diagram[1].

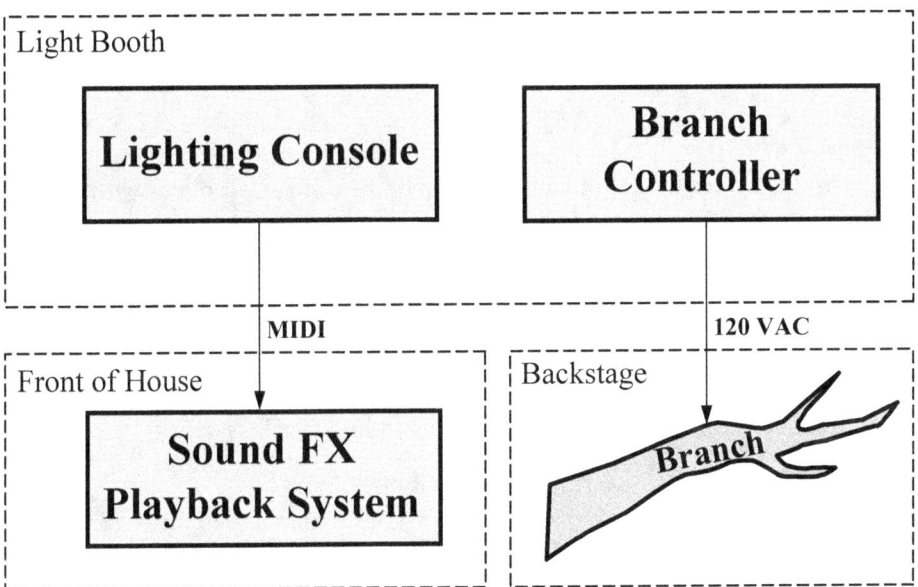

For the test, the four lightning/thunderclap effects are triggered by the electrician, using the console's fader buttons, on cues from the stage manager. The MIDI output from the lighting console is routed to the MIDI input of the sound playback system, so that pressing a fader button triggers both the lightning effect and the appropriate sound. So that the delay between lightning flash and thunderclap will be consistent night after night, the sound designer programs a "Wait" cue into the sound playback system, which, when triggered, automatically delays the thunderclap sounds by the appropriate amount. Once the stage manager confirms that the tree effect has been armed, they call the cue for the falling branch over the headset to the electrician; the sound operator is cued by a cue light for the falling-branch sound. The breaking-glass sound is taken as a visual cue by the sound operator as they see the tree poke through the window.

The director and design team are delighted with the results of the test, but there are a few problems. The electrician is not able to consistently control the duration of the lightning flashes to the director's and lighting designer's satisfaction, and the director now feels that the timing between the first three thunderclaps should be exactly the same every night, regardless of what the actors are doing (in other

---

1 Note, the light booth cable run is more than 50 feet, so the sound department makes some MIDI DIN to XLR adapters and runs through some XLR tie lines. The sound department tests this thoroughly to verify that the connection is solid.

words, they want to "can" the effect). In addition, during the first test, the big thunderclap was so loud that the electrician never heard the "Tree Release Go" command and missed the cue. Finally, the director thought that the lightning for the big blast was not bright enough, so an additional strobe is rented and connected to the lighting console.

Approach 2

The sound designer figures out a way to solve many of the problems with the first approach. The sound playback system can, in addition to generating sound, also generate any sort of MIDI message. So, they program in MIDI Show Control messages for the light board, removes the wait time from the start of each thunderclap cue, and sets the delay between the lightning flashes and the thunderclaps using a time clock available in the sound playback system. The MIDI lines in place are reversed, sending data from the sound playback system to the lighting console.

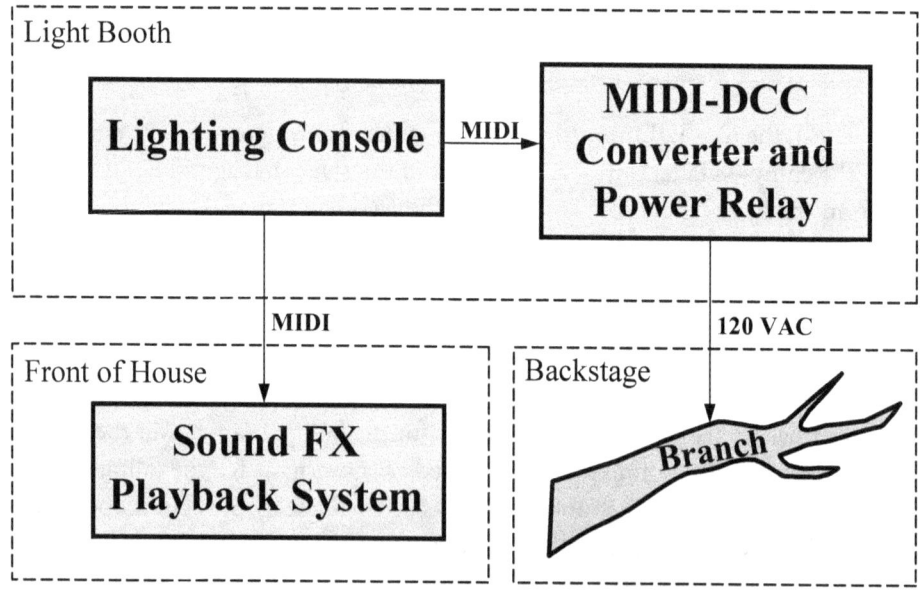

To fire the tree effect reliably, the electrician buys an inexpensive MIDI-to-dry-contact-closure (DCC) relay interface, which can fire the solenoid using a MIDI Show Control machinery commands: "Go" is programmed to turn the solenoid on; "Stop" turns it off. The relay interface will be controlled by the sound playback system, and since MIDI is already run from the sound playback system to the light board backstage, the relay interface is simply placed next to the console and given a MIDI input. The sound designer programs two sequences into the sound

FX playback system that the sound operator will execute with single button presses. Sequence 1 contains the messages for the first three lightning strikes (Note: the "Time Code" column is the system's internal clock):

Time Code	Cue	Effect	MIDI Command
00:00:00:00	218	Lightning Strike #1	MSC Cue 218 Go
00:00:02:00	101	Thunderclap 1	
00:00:15:00	219	Lightning Strike #2	MSC Lights 219 Go
00:00:16:00	102	Thunderclap 2	
00:00:26:00	220	Lightning Strike #3	MSC Lights 220 Go
00:00:26:05	103	Thunderclap 3	

Sequence 2 consists of the simultaneous lightning effect and huge thunderclap, followed by the release command for the tree and the falling-tree sound effect:

Time Code	Cue	Effect	MIDI Command
00:00:00:00	220.1	Big Lightning Strike	
00:00:00:00	110	Big Thunderclap	
00:00:03:00		Tree Release	MSC Machinery Go
00:00:04:00		Tree Release Reset	MSC Machinery Stop
00:00:04:00	111	Tree Falling Sound	

These two sequences allow the entire thunderstorm to be run with two simple Go commands. Since the rate at which the tree branch falls is not predictable, the breaking-glass sound cue is still taken visually by the sound operator. The final "lights out" that ends the act is called by the stage manager and taken manually by the electrician.

This new approach works well, but the additional strobe was mistakenly plugged into the same power circuit as the mixer and generated a huge static blast through the sound system, and the sound operator was so busy pulling down channels on the board to kill the static they were late in executing the second, tree-falling sequence. Flustered, the operator missed the breaking-glass sound altogether.

Approach 3

To overcome all these problems, the team decides to use even more of the show control functionality of the sound playback system and use the movement of the tree itself to trigger its sound effect. This leaves only the problem of triggering the breaking-glass sound effect. Since the sound operator is no longer running the rest of the sequence, they are free to run the cue manually, but could be tied up at any point with unanticipated live sound reinforcement problems. So, the sound designer decides to trigger the sound automatically.

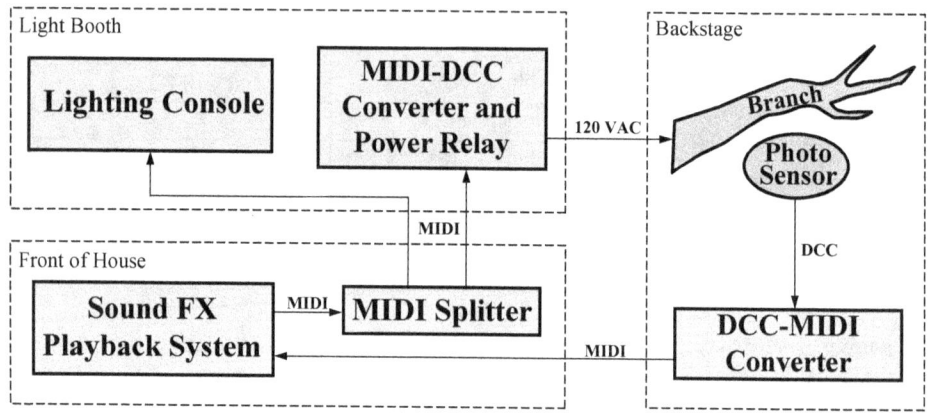

The tree does not fall at a predictable rate, so the sound department installs a photo-electric sensor across the window frame; when the falling tree breaks the beam of light, the photo sensor generates a contact closure. A DCC-MIDI interface is programmed so that a MIDI Show Control command is generated when the beam of light is interrupted (another is generated when the beam is cleared). The cue can also be triggered manually if there is a problem with the sensor.

This approach works beautifully, the director is ecstatic, the show gets great reviews, and sells out.

TEN-PIN ALLEY

The animatronic bowling ball raises up and beckons the crowd, "Hey kids, come over here and watch an amazing show with me, Bowly!" The house lights dim and moving lights fill the space with color. "You, too, grown ups," says Bowly. A catchy bowling tune replaces the store's background music and sound effects, and mapped video is projected on the columns around the store while the main video starts playing on a huge video wall at the back of the space. "Here we go!" shouts Bowly, and video images of dancing bowling pins fill the screen, cut precisely to the music beat. At the end of the five-minute show, the house lights and regular background music restore, and Bowly says, "Hey, thanks for coming, the next show will be in a few minutes. In the meantime, why not come inside and take a look around?" Video monitors around the floor, which had been relaying a feed from the show, spring back to life and display live bowling highlights from around the world, intercut with real-time score updates from the adjacent bowling alley.

The Mission

"Kingpin," president of 300 Score Entertainment (3SE), wants this sequence to take place every 30 minutes at StrikeTown bowling superstores/alleys around the world. The flashy shows are designed to attract customers into the store, get them to stay a little longer, and maybe buy some StrikeTown sportswear or a new bowling ball, and later bowl a round, or dine and dance in the attached bar/restaurant complex.

Kingpin has decided to build the flagship StrikeTown in Las Vegas, and has hired a design firm to produce the shows, and the firm will be producing the video and audio media. The design firm's producer/director is working with Kingpin's architect for scenic elements, and has hired big-name Las Vegas lighting and sound designers to create the lighting and sound in the store. Freelance lighting, sound, animatronic, and show control programmers are hired to program the show.

Kingpin has hired a general contractor (GC); the GC has hired an entertainment lighting contractor to handle entertainment lighting systems; a scenery automation company to handle the animatronic lift; an A/V contractor to handle audio, video, and control; and an animatronic character company to provide Bowly and its control system. On the A/V contractor's recommendation, Kingpin has hired a show control consultant to coordinate the control of all the systems.

Shows will run automatically from a centralized "Scheduler" computer, which can be updated from corporate headquarters in Milwaukee over the corporate intranet, or from the store manager's desktop. Shows can also be run manually from a touch screen controller for special events, such as weddings, or appearances by famous bowlers with StrikeTown endorsements.

3SE bean-counters will allow only one maintenance person for all the technical systems in the store and the bowling alley, and this person will be responsible for everything from the show systems, to a fancy glass elevator to bring bowling balls up from the basement warehouse area, to the pin re-racking equipment. The system will have both a "day" mode when the store is open and a "night" mode when the store is closed and show equipment is shut down.

What are the safety considerations?

The large screen and the animatronic character "Bowly" are potentially dangerous, but the audience is physically partitioned away from each and interlocks can be placed on the access points.

What type of show is it?
It's a linear show, since it's all prerecorded.

What kind of triggers and synchronization should be used?
The show, since it's all synchronous, will use some sort of time code.

What devices or systems are to be connected/controlled?
Let's go through each system.

Scheduler
The Scheduler is going to be created, installed, and maintained by 3SE's Information Technologies (IT) department, since it is going to run on the corporate intranet. The IT department originally wanted to oversee all the control aspects of the show and store, but after hearing some war stories from the show control consultant, they decided to let the show production contractors handle it. So a separate system, specified by the show control consultant, will run the shows and be triggered at the appropriate time over the network by the Scheduler. The Scheduler will maintain a web page on the corporate intranet for remote monitoring, and will offer web-cam views of the store and backstage equipment room.

The Scheduler will send a variety of ASCII commands via the network to the show controller. One minute before show time, the Scheduler will send a "Warning Show #n" command and later a "Go Show #n" message to start the show. Each command will be acknowledged by the Show Controller, which will control everything associated with the show. Both the Scheduler and Show Controller will be able to communicate with various store systems.

Show Video
Video is a key element of the shows. To ensure maximum flexibility and to allow shows to be easily updated, video servers can have their contents updated over the corporate intranet. The control connection to the server is via Ethernet, using proprietary ASCII commands. The server also has a balanced XLR time code input and output, and can chase or generate time code.

Each video segment is assigned a different time code hour, and one minute of video black is included at the head of the segment before the actual show starts so that "Bowly" can start and run and other systems can roll up and set up into a preshow state. One minute of video black fills out the end of each segment. Here are the details of two typical video segments:

Time Code	Event
11:00:00:00	1 minute pre-roll video black
11:01:00:00	Start of "Bowling's Greatest Bloopers" segment
11:05:04:25	End of bowling segment
11:06:04:25	End of video black
12:00:00:00	1 minute pre-roll video black
12:01:00:00	Start of "King Pin" segment
12:06:01:00	End of "King Pin" segment
12:07:01:00	End of video black

The show video projectors also need some control so that they can be configured, put into or taken out of "standby" mode at the start and end of each day, and queried before shows. The projectors have Ethernet ports and accept and respond with proprietary ASCII commands.

Store Video

The store video systems have been contracted separately, and video for the 100 or so monitors around the facility is sent through a routing switcher, which is controllable via Ethernet and ASCII commands. The switcher can route any of a number of satellite-received sports channels to the various monitors, put up logos, take graphics feeds for messages or score highlights from the alley, or take the video feed from the shows.

Show Audio

Audio will be played off a networked audio playback system which can chase SMPTE Time Code or accept ASCII trigger commands over the network. The audio mix for the store will be accomplished through a computerized mixing matrix system, which can accept Open Sound Control (OSC). All the amplifiers for the store are connected, monitored, and controlled from a control system made by the amplifier manufacturer, which also can talk via a proprietary link over the network.

Store Background Audio

The contract for store audio is separate from the show audio, and while similar equipment is used in some parts of the system, a matrix system designed for permanent installations (rather than shows) is used. It too can respond to simple

ASCII commands over Ethernet.

Animatronics

"Bowly" has its own control system, which can chase SMPTE or MIDI time code, or be triggered by commands over a network or contact closures. The consultant gets the manufacturer to add Ethernet support to keep the whole system consistent. A programming panel is provided for inputting Bowly's moves, along with a small animatronic control rack, so that Bowly can be programmed "offline" separately from the show and then integrated when its programming is complete. Photo sensors are used to create a perimeter around the character; if a customer gets too close the animatronic system will go into E-Stop mode.

Stage Machinery

The animatronic lift motion-control system takes commands over the network to set its target position. A manual control panel with auto/manual lockout is provided, as is a complete E-Stop system. The lift is housed in a vertical area closed to the public, accessible only via a locked door. For safety reasons, a sensor on the door sends a signal to both the motion-control system and to the Show Controller (for status reporting) each time the door is opened. When the sensor is tripped, the lift, if moving, will stop, and the system will have to be manually reset using a key-switch near the door before the screen will be allowed to move again.

Show Lighting

The show lighting system controller is available in both a console version with full controls and a rack-mount version. A full console is rented for show programming, and the show is then loaded into the rack-mount system for daily operation. The controller can chase SMPTE or MIDI Time Code, or accept proprietary ASCII commands over Ethernet.

Store Lighting

The contract for store lighting has been let separately from show lighting. The store system is fairly simple, with two general "looks": normal and show. The store controller can accept ASCII commands over the network, and also acknowledge via a simple "Ack" for the pre-show test.

What system architecture works for this application?

Because of the size and complexity of the system, a show controller is needed, and will be connected to other devices mostly over a network in a peer-to-peer way to allows maximum flexibility.

What is the control information source?

The Scheduler will start each show, and will be responsible for scheduling day and night modes. In addition, the store manager will be able to override certain shows or force a "party" mode.

What type of user interface is required?

There is only one technician on the premises, so the system should present them with a detailed interface, but should also present a simple system to the manager for overrides, parties, and so on.

What kind of show control approach is needed?

The consultant has a lot of experience with a hardware-based system, and has suggested that for this application. This system can generate time code; chase it; send serial, MIDI, or other messages; send and receive any sort of messages over Ethernet; and send or receive contact closures wired to the back panel of the unit. The show control consultant decides as much interconnection as possible should be done via Ethernet, and they lay out the IP addresses in consultation with the company's IT department (the show network is separate, but by coordinating IP addresses, there's room for the future and there won't be a problem if the systems are accidentally connected). At the one-minute warning, the Scheduler sends its control message via the network to the Show Controller. The Show Controller then checks each connection, sending various "Are you online?" commands to each of the subsystems, and waits for replies from each. If all the critical systems check OK, the controller sends a message back to the Scheduler saying that the show is standing by, and this is logged on the scheduler's Web page. If the systems do not respond properly, the Show Controller sends a message to the Scheduler detailing the problem, and the Scheduler can put this information on its Web page, send an e-mail, or even page maintenance personnel.

The consultant creates a system block diagram and user screen:[2]

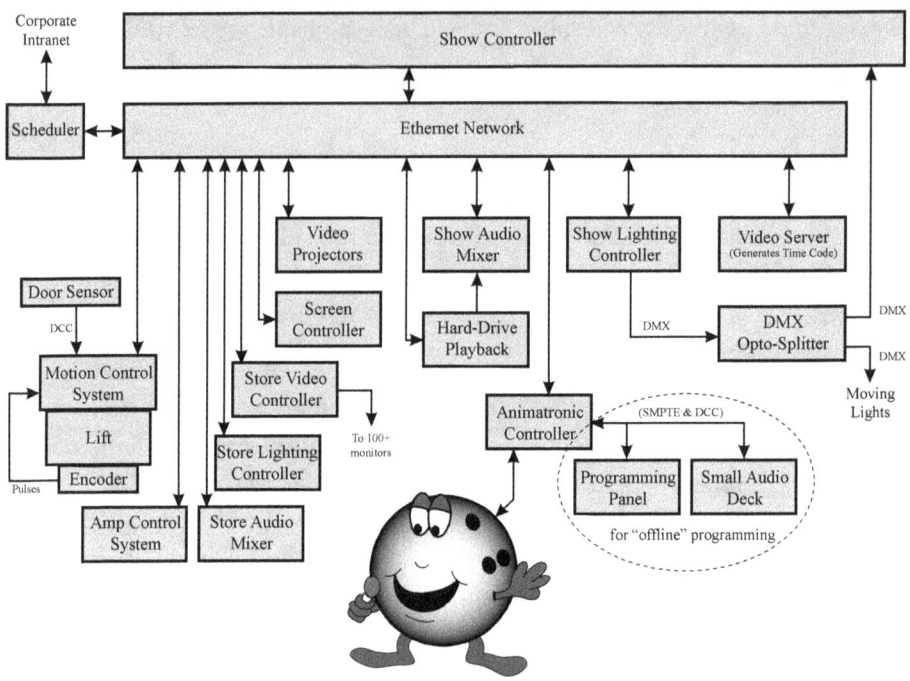

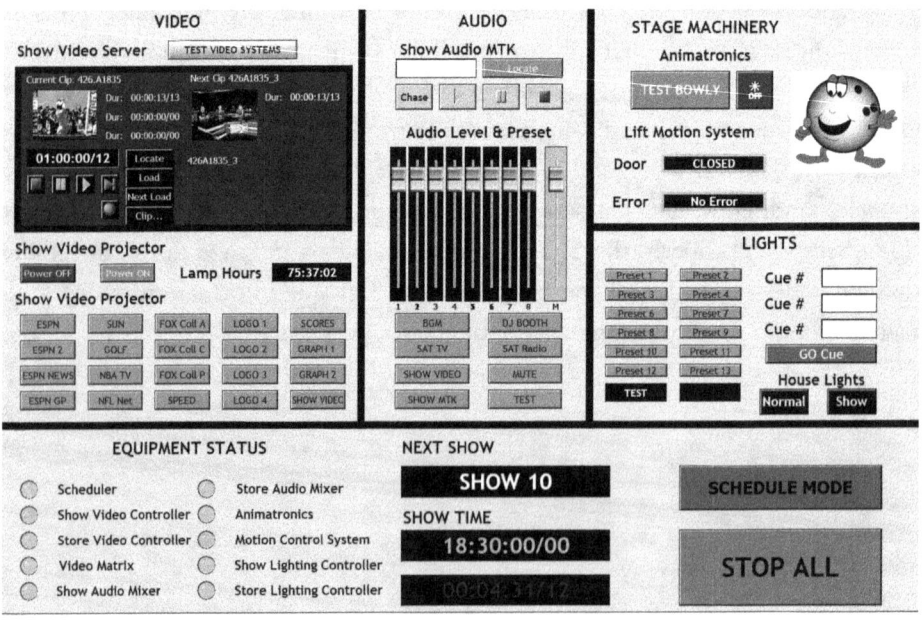

---

2 User screen by Alan Anderson.

The show control consultant generates a show control script, one segment of which is shown below.

Trigger	Event	Action
One minute before show time	One minute warning	Scheduler sends message to Show Controller (SC)
SC receives message		Start Show Pretest
Show pretest	SC tests lift	SC sends message to lift system
Lift replies	Lift OK or not OK	SC sends "Lift OK" (or "show abort") message to scheduler
Lift test complete	SC tests Show Video	SC sends message to video server SC sends message to projectors
Video server/ projectors reply	Video/projectors OK or not OK	SC sends "Show Video OK" (or "show abort") message to scheduler
Show video test complete	SC tests Store Video	SC sends command to store video
Store Video replies	Store Video OK or not OK	SC sends "Store Video OK" (or "not OK") message to scheduler
Store Video test complete	SC tests Show Audio	SC sends message to show sound matrix, SC sends message to amp controller
Matrix and amp controller reply	Matrix/Amp Controller OK or not OK	SC sends "Show Audio OK" (or "show abort") message to scheduler
Show Audio test complete	SC tests Store Audio	SC sends message to store audio
Store Audio replies	Store Audio OK or not OK	SC sends "Store Audio OK" (or "not OK") message to scheduler
Store Audio test complete	SC tests Animatronics	SC sends message to animatronic controller
Animatronic Controller replies	Animatronics OK or not OK	SC sends "Animatronics OK" (or "not OK") message to scheduler
Animatronic Test complete	SC tests Lighting	SC sends message to show lighting
Show Lighting replies	Show Lighting contact closure on or off	SC sends "Show Lighting OK" (or "not OK") message to scheduler
Show Lighting Test complete	SC tests Store Lighting	SC sends message to store lighting system
Store Lighting Test complete	SC evaluates if critical systems (Screen, Show Video, Show Audio) are OK	SC sends "Standing By for Show" message to Scheduler or sends "Critical System Failure, Show Aborting," and cancels show
Show Time	Show 12 Go	SC starts Video Segment 14
Start of show	Video starts rolling	Time Code is generated
12:00:00:00	Time Code rolls	

Trigger	Event	Action
12:01:00:00	Bowly starts talking	TC triggers animatronics preshow
12:04:00:00	Store audio starts to fade down	SC sends message to store audio
12:04:30:00	Store lighting starts to fade down	SC sends message to store lighting
12:04:55:00	Lift raises	SC sends message to lift system
12:04:58:00	Bowly says "Here we go!"	
12:05:00:00	Show starts	
Various	Show lighting/sound chase time code	
12:10:01:00	Show ends	
12:10:02:00	Lift retracts	SC sends message to lift system
12:10:06:00	Store lighting starts to fade up	SC sends message to store lighting
12:10:07:00	Store audio starts to fade up	SC sends message to store audio
12:10:10:00	Bowly finishes speaking	SC sends "Rest" command to Bowly
12:11:06:00	End of time code	SC sends "Show Complete" message to Scheduler, and systems reset as necessary for next show

The flagship StrikeTown is a huge success. The only problem is that the maintenance issues are too much for one person to handle, so a second full-time person, who deals only with show systems, store audio, and store video, is added. Kingpin announces future StrikeTown locations in Orlando and Abu Dhabi.

COMFORTABLY RICH

During "Comfortably Rich," the band's biggest hit, images of a flying poodle, edited perfectly to the song's beat, are showing on a giant circular screen. Video walls built into the set switch back and forth from prerecorded poodles to live views of the band. Throughout the song, sound effects and barks from the poodle footage emanate from the massive quad sound system. Then, suddenly, the drummer's riser lifts 20 feet into the air, with massive spots tracking the riser from every direction; this is the moment the crowd has been waiting for—the drum solo. As the riser's ascent slows, the poodle images disappear and the solo starts. As the drummer hits each of their drums, the huge array of moving lights dances precisely to the beat. At the end of the solo, the drummer plays the famous opening beats from "Young Rust," and as the drum riser returns slowly to earth, images of rust and money appear on the screens.

The Mission
Tribute band Purple Floyd want this spectacle recreated nightly for their final tour (the tribute band had its first "final" tour five years ago).

What are the safety considerations?
The movement of the mechanized drum riser is potentially dangerous, and this needs to be taken into account.

What type of show is it?
It's mostly a linear show, since the band will play (more or less) the same sequence of songs each night.

What kind of triggers and synchronization should be used?
Some sort of time code will likely be needed in conjunction with the media, but there will probably be a few event-based triggers as well.

What devices or systems are to be connected/controlled?
The band's longtime production manager is in charge of contracting the various people and companies that will supply and operate gear for the tour. Since this is the first time the band is trying this kind of integration, the production manager decides to bring in a show control consultant to ensure that the critical sequences work correctly.

Lighting
The moving-light console has a MIDI port which can accept MIDI Show Control commands; it can also accept ASCII console commands over Ethernet. The following looks have been programmed into the console for the drum solo:

Drum/Note Number	Control Message
Kick/C3	Go Cue 1, List 10
Snare/D3	Go Cue 2, List 10
High Tom/E3	Go Cue 3, List 10
Mid Tom/F3	Go Cue 4, List 10
Low Tom/G3	Go Cue 5, List 10

Sound

Purple Floyd's quad sound system is controlled using a large digital console, and uses a network audio distribution system. A fiber-optic network-based snake, with a redundant back up system, has been run; it can carry audio, SMPTE, and MIDI through the use of various interfaces. Audio for the video segments is provided by the video department.

Video

Live video from the cameras and playback sources is controlled by a video director located backstage with all the video gear. Video servers provide high-definition video to the projectors and video walls over a high bandwidth video network, switching back and forth with the live feeds. The video servers can accept DMX or sACN control, and also chase time code.

The poodle video segment is six minutes, 23 seconds, and four frames long; the rust segment is three minutes, 25 seconds, and 20 frames long. Each prerecorded video segment in the show will be assigned a different time code hour: The poodle segment is the third in the show, so it uses hour 3; the rust segment is next, so it uses hour 4.

To make sure everything can sync up, 10 seconds of video black is added on to the head of each video segment and onto the tail (end) of each segment, to fill out to the next largest time-code minute. Here's an example from the poodle segment:

Time Code	Event
03:00:00:00	10 seconds preroll video black
03:00:10:00	Start of poodle segment
03:06:33:04	End of poodle segment
03:07:00:00	End of video black
04:00:00:00	10 seconds pre-roll video black
04:00:10:00	Start of rust segment
04:03:35:20	End of rust segment
04:04:00:00	End of video black

Stage Machinery

The drum riser will lift to a preset position on receipt of an ASCII string "Up" over an Ethernet connection; a "Down" string tells the lift to retract. A special "enable" button, which is downstream of any network control, is located near the lift control. The backline drum technician will press and hold the authorize button when the drummer and all crew are in a safe position, enabling the effect. Finally, the Stage Machinery system can broadcast its current position over a network every 100 ms.

Musical Synchronization

The band and the crew feel most comfortable putting everything on time code. However, for everything to work seamlessly, the musicians somehow must lock themselves to the time code related to the media, so the sound engineer creates a click track which is played from the video server to the drummer's in-ear monitors. Once synced to a video segment, the drummer and the rest of the band can "float" a little bit within the timing of the segment, as long as they mostly stay in sync. To make the first transition from a time-based to a spontaneous segment, the drummer simply begins their solo as the "Comfortably Rich" video segment ends. When the drummer wants to end their solo, they signal the operators to start the next segment by playing a special beat pattern, which is never played except at the end of the solo. The backstage operators roll the next segment, and the drummer is able to resync their tempo (and therefore the band) by listening to the click track. A similar approach is taken to get into and out of guitar solos and the like.

To synchronize the lighting with the electronic drum kit, MIDI Note messages are taken from the drum set and must be processed to trigger corresponding MIDI Show Control Messages.

What system architecture works for this application?
Because of the size and complexity of the system, the consultant wants a show controller which will be connected to other devices mostly over a network in a peer-to-peer way to allows maximum flexibility. But since their is a master time clock, the main part of the song will run in a primary/secondary way, with the subsystems chasing the clock.

What is the control information source?
In this case, it is the band itself.

What type of user interface is required?
All the system operators on the tour are skilled technicians, so they want very sophisticated and complete user interfaces.

What kind of show control approach is needed?
While the video cue lists tell part of the synchronization story, the whole sequence is getting quite complicated now, so the production manager generates a script.

Trigger	Event	Action
Manual	Comfortably Rich preroll	Start video segment TC: 03:00:00:00
8 beats before first frame	Start of click track	Click track to drummer
03:00:10:00	Comfortably Rich segment start	
03:05:00:00 (Approx)	Check riser safety	Enable riser movement (Drum Tech)
03:06:15:00	Drum riser rises	
03:06:33:04	End of song/start of solo	
Manual	Enable MIDI drum lights	Trigger light looks
03:06:35:00	Cue video for Young Rust	
Manual	Disable MIDI drum lights	
Drummer's special beat pattern	Young Rust Pre-Roll	Start video segment TC: 04:00:00:00
Manual	Lower riser	Disable riser movement (Drum Tech)
8 beats before first frame	Start of click track	Click track to drummer
04:00:10:00	Young Rust segment start	
04:03:35:20	End of song/segment	

The band has used the same video director for years, so they decide that they will be the one who pushes the "play" button, starting the video segments at the correct times. When the button is pressed, SMPTE Time Code is sent out through a time code distributor/analyzer to the lighting console, so that it can be cued tightly with the video; to the sound console, which does some automated panning; and to a show control system, which sends out network commands to the drum riser.

To get the drum solo lighting to trigger, the consultant configures the show controller to generate appropriate messages over the network for the lighting console when it receives corresponding MIDI note messages from the drum kit.

The show control consultant creates this block diagram and user screen[3]:

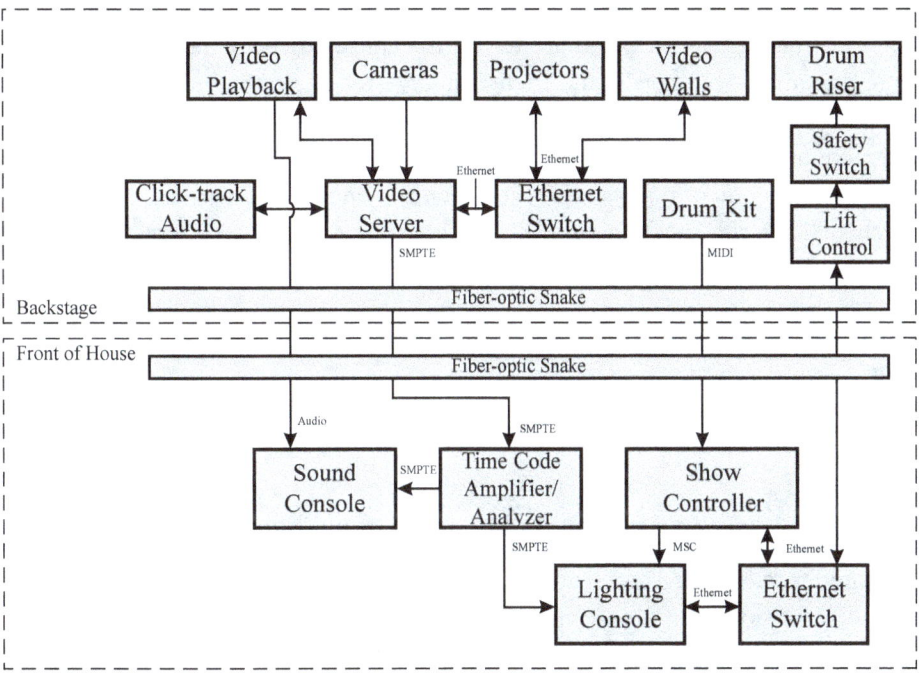

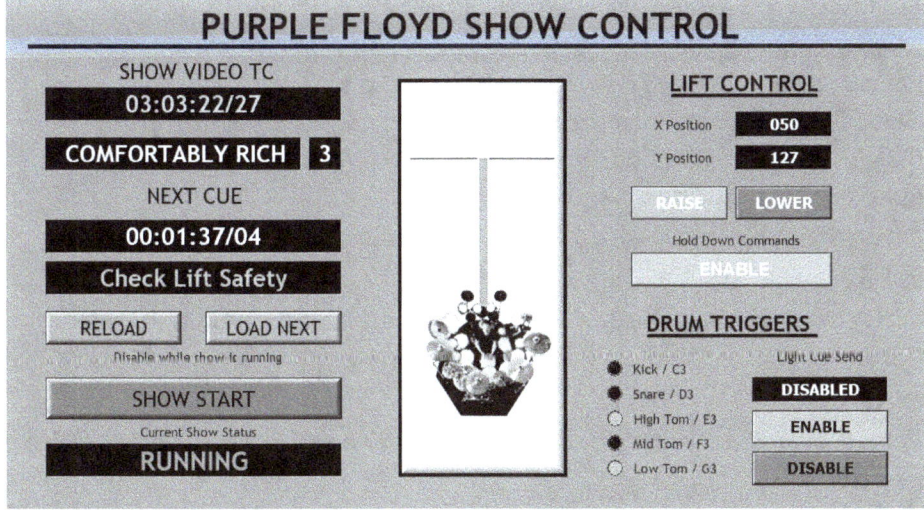

The production manager wisely arranges a large-scale test about two months before the tour is to begin rehearsals. The band loves it, and the tour is a massive success!

---

3 User screen by Alan Anderson.

POPSICLE WOLF

The audience wanders in, and are somewhat confused. They have entered into a large unmarked warehouse, and there is a large robotic cat named Popsicle looking out from its perch atop a huge, oversized cat tower. But that's it—no further instructions are given. Eventually, one of the audience members is brave enough to go through one of the oversize cat doors, and encounters a series of connected, psychedelic worlds; one is a hall full of mirrors; in another, an interactive musical instrument fills an entire room; in yet another is like a museum of artifacts, and further still is a room is Popsicle's "Purratorium". Costumed characters occasionally appear and lead audience members into secret experiences. Exploring, the audience discovers thirteen worlds, and digging deeper they see clues and discover a story threaded throughout the massive attraction.

The Mission

This is what the creators have envisioned for the new Popsicle Wolf immersive art experience. With so much going on at once, it's nearly impossible to ever see it all, and it would take multiple visits to really follow all of Popsicle's story. Popsicle Wolf started out as a crazy art installation and now has grown to multiple installations, each one unique. Popsicle Wolf now has an in-house show control team, and they are designing this new attraction in Fargo.

What are the safety considerations?

The risks in this attraction are mostly that the audience might trip over something in the maze-like attraction, but none of the connected systems are planned to have any intrinsic risks.

What type of show is it?

Since the audience can walk wherever they want, this attraction is definitely non linear.

What kind of triggers and synchronization should be used?

With the structure of the experience, the design team has decided to use a hybrid approach, where as the audience passes by or triggers specific sensors in each of the worlds, a short timed sequence will be triggered.

What devices or systems are to be connected/controlled?

A wide array of devices will make up this show, distributed between the lobby and the thirteen worlds. Let's look at each department.

Lighting

To keep things manageable, there will be a large, centralized lighting control system which sends signals out over the show network to all the lighting around the attraction. Multiple people can be logged in and working in different areas of the attraction at the same time. This system can talk over a network either via OSC or a manufacturer specific network protocol.

Sound

A centralized sound control system sends audio signals out over the show network to distributed speakers around the attraction. Multiple people can be logged in and working in different areas of the attraction at the same time. This system can talk over a network either via OSC or a manufacturer specific network protocol.

Video

A centralized video playback sends video signals out over the show network to distributed screens and projectors the attraction. Multiple people can be logged in and working in different areas of the attraction at the same time. This system can talk over a network either via OSC or a manufacturer specific network protocol.

Sensors and Audience Interactive Controls

To make wiring and management of this sprawling attraction manageable, the inputs to the show control system are distributed out throughout the attraction. In each world and the lobby, an input box is positioned that takes contact closures from sensors and audience controls (pushbuttons, etc) and then sends that information to the show control system over the network.

Effects

There are a number of small mechanical and special effects also distributed throughout the thirteen worlds. To facilitate wiring and safety interlocks (where needed), an effects control cabinet is placed in each of the worlds. This communicates to the show control system over the network via a simple proprietary protocol.

Popsicle Kitteh Animatronic

The animatronic character in the lobby has its own dedicated control system also connected via the network and communicating via a proprietary protocol developed by the animatronic control manufacturer.

Remote Access

The team at Popsicle Wolf headquarters in Santa Fe want to be able to monitor and control the system remotely. For this reason, one computer is added in the main equipment rack which has two network jacks: one on the internal show network and the other connected to the Internet. A special firewall is added to the connection, and the machine has multiple security precautions implemented and uses a tightly controlled, logged, encrypted remote access solution.

What system architecture works for this application?
Given the scale of the attraction and the needs of various departments, the show control designers decide on a distributed peer-to-peer architecture for the system.

What is the control information source?
In this case there are several sources of control information: sensors that the audience trigger (without knowing it), various controls and buttons that they may interact with, and then the staff user screens. In addition, there will likely be some show aspects triggered at certain hours of each day.

What type of user interface is required?
While not apparent to the audience, there are a number of operator control touch screens hidden and locked away throughout the attraction; one for each world, plus additional screens where needed, such as a status monitoring screen in the equipment room. Here's the touch screen[4] for the Purratorium:

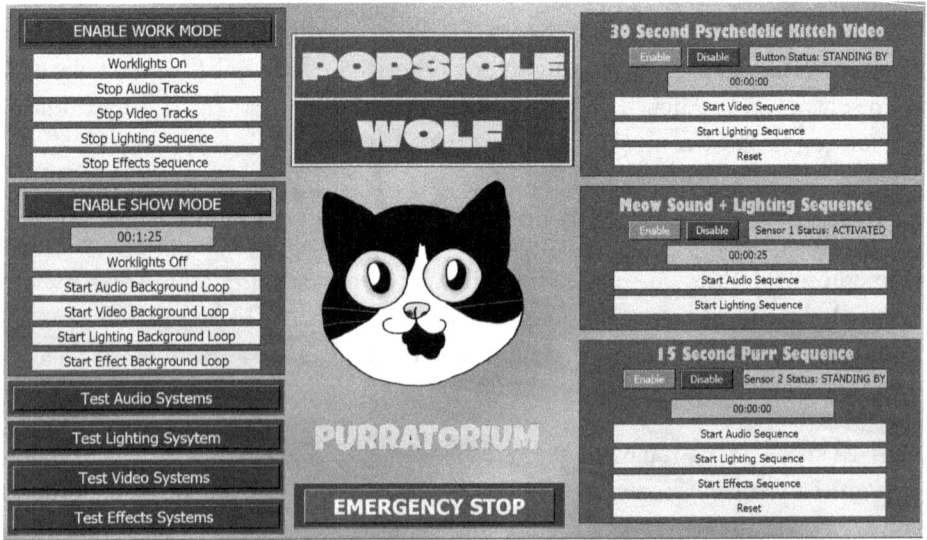

4 User screen by Daniela Tomaino

What kind of show control approach is needed?

The show control team comes up with the following signal flow diagram (some areas omitted for clarity):

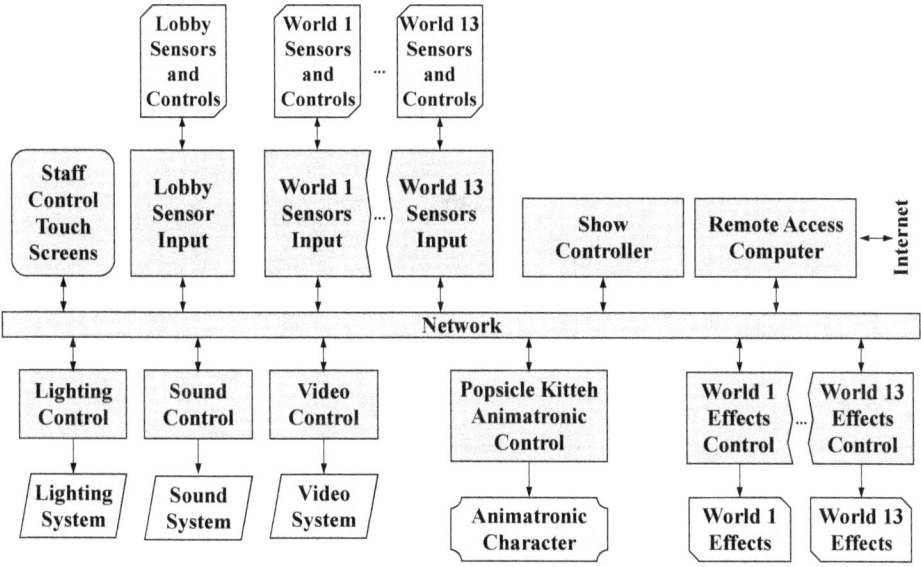

All the gear is in an equipment room, and there is a main touch screen in the lobby where the attraction is started at the beginning of the day and shut down at the end. On start of show background audio/music, lighting, video and effect loops are triggered, and sensors throughout the attraction are enabled. (This screen also contains maintenance and troubleshooting screens).

A cue list is developed for each area. Let's take a look at a simplified excerpt of the the list for Popsicle's Purratorium.

Trigger	Event	Action

Button press on lobby user screen	Show Mode	`Go Lighting WL-000` (Worklights off)
		`Go Audio PP-001 Background Loop`
		`Go Video PP-001 Background Loop`
		`Go Lighting PP-001 Background Loop`
		`Go Effects PP-001 Background Loop`

Audience passes sensor 1	Meow sound and lighting sequence	`Go Audio PP-011 Sequence` `Go Lighting PP-011 Sequence`

INTRODUCTION TO SHOW CONTROL • 93

Trigger	Event	Action

Audience pushes button	30 Second Psychedelic Kitteh video	`Go Video PP-101 Sequence`
		`Go Lighting PP-101 Sequence`
		`After 30 Seconds`
		`Go Lighting PP-111 Sequence (Reset)`

Audience passes sensor 2	15 second purr storm	`Go Audio PP-201 Sequence`
		`Go Lighting PP-201 Sequence`
		`Go Effects PP-201 Sequence`
		`After 15 Seconds`
		`Go Lighting PP-211 Sequence (Reset)`

Button press on lobby user screen	Work Mode	`Go Audio PP-900 Background Stop`
		`Go Video PP-900 Background Stop`
		`Go Lighting PP-900 Background Stop`
		`Go Lighting WL-001` (Worklights on)

The Fargo Popsicle Wolf experience is such a hit that they decide to replicate it for the first ever Popsicle Wolf experience in Japan.

IT'S AN ITCHY WORLD AFTER ALL

Itchy and Scratchy wander on to the huge outdoor stage from opposite ends. Itchy, proclaiming his friendship, gives a bouquet of flowers to Scratchy, who is purringly happy. But, seconds later, Scratchy gives his trademark scream as Itchy shoots him off the stage with a large-caliber machine gun. Scratchy staggers back onstage to get a drink of water from a well, but Itchy sneaks in from the side and kicks Scratchy down the opening. The manic mouse then throws a hand grenade down after the crazy cat, and a huge explosion bellows out, spraying the audience with water. Moments later, a charred Scratchy appears at the mouth of the well, staggers across the stage, and lays down under a ledge to get a badly needed cat nap; little does Scratchy know that this is actually a bowling alley. Itchy, however, is happy to point this out, and quietly arranges nine warhead bowling pins around the napping cat. Next, Itchy rolls a bomb/bowling ball down the alley and there is a huge explosion. "Cut" yells the "director," as the smoke clears, and the Itchy and Scratchy performers come back onstage. "Now, audience," the director asks, "what should we do next?"

The Mission

These are the first three segments of the *Itchy and Scratchy Epic Stunt Spectacular*, which will purportedly show how action movies are made, using indestructible cartoon characters Itchy and Scratchy to demonstrate various stunts. This is to be the main attraction at Duff Gardens, the internationally known theme park located near Orlando, Florida. Unlike most theme park shows, this one has a twist: The audience will select which stunts they want to see, and in which order.

Mr. Burns, Duff's CEO, has taken a personal interest in the show, and so everyone is working hard to ensure that everything goes smoothly. Itchy and Scratchy are Duff's corporate icons, so no expense is spared. However, while the attraction's construction is lavish, its operating costs will be kept as low as possible, because the park management intends the show to run for many years. So, the entire show will be run by one Technical Director (TD), one Sound Operator, and several pyro and special effect technicians.

What are the safety considerations?

There are many dangerous effects in this show, so all proper precautions must be taken.

What type of show is it?

It's a nonlinear show, since the audience selects which stunts they want to see and in which order.

What kind of triggers and synchronization should be used?

The show will be a loose, event-driven collection of flexible time-based sequences.

What devices or systems are to be connected/controlled?

Equipment for the production is being supplied by a number of subcontractors, all working for DGI (Duff Gardens Illusions), Duff's in-house design and engineering department, which is designing and providing general contracting services for the show.

Lighting

The lighting system will be built around a rack-mounted version of a large control console. A full-sized console will be used for the programming period and then the show will be downloaded into the rack-mounted unit for show operation. Either version of the console can be driven via commands over the network, that will also be used to connect all the conventional dimmers, moving lights, and LED fixtures. The network also accepts commands from a special processor which provides interfacing for emergency lighting systems, allowing certain looks to be brought up independently of the show control system in the event of a park-wide or other emergency.

Video

High-power video projectors are used to generate massive images for the nighttime shows, projecting pre-show images on a nearby water tower shaped like Itchy's mouse ears. Two other video outputs drive daylight-viewable LED screens for sequence titles, voting results, and other show imagery. Each projector is driven by a video server, which is connected to the show control system via Ethernet.

Sound

The Sound Operator operates an audio mixer and a sound FX playback triggering system—both of which can accept various control messages over Ethernet. The mixer handles all routing and gain manipulation for the system, and receives audio inputs from the sound FX playback system, and several wireless mics. In addition, it can take commands over the network from the show control system, as well as from park-wide management. A park-wide connection enables special emergency paging lines to be made active at any time and incorporates park background music and a general paging system.

The performers playing the characters do not actually speak; instead, each individual phrase (or scream) is triggered manually from the sound FX playback system by the Sound Operator using a push button panel. This approach ensures proper synchronization regardless of what happens onstage, and in rehearsal, the actors learn certain moves to indicate to the operator when to trigger the sound. This system also plays back other prerecorded sounds, such as the voice-over used to open the show, which are triggered from the show control system over the network.

Effects

There are extensive pyrotechnic and other effects in this production, which will be controlled using a computer-based pyro controller that incorporates a number of safety systems. The safety system determines whether the effect is safe to go based on the state of a variety of detectors and, most importantly, TD and actor-operated safety switches. For dangerous effects, in addition to the TD "authorizing" the effect, the performer must press and hold a button or buttons, indicating that they are is in a safe position, or the effect will not be allowed to operate. Until the effect is safe to go, the firing contacts to the actual effect are shorted by the safety system, ensuring that no matter what happens upstream in the pyro controller, the effect cannot fire until all is safe. An Ethernet connection from the safety system is used to report status back to the show controller. For safety reasons (and to save money), pyro substitutes are used whenever possible; the substitutes are also fired by the pyro control system because some of them still present a danger to the performers. The well explosion, for example, is really a pneumatic "air cannon" effect, which blasts a tremendous amount of water out of the well in conjunction with a sound effect. The machine-gun bullet hits are also pneumatic.

The machine-gun effect uses a special system to give the performer maximum flexibility, and to make the effect as believable as possible, it enables the performer to actually control the machine-gun effects directly. When the trigger is squeezed on Itchy's machine gun, a radio signal is sent to a receiver, which generates a message over Ethernet to the show control system. When enabled by the TD through show control, the pneumatic bullet-hit effects are then triggered, along with localized machine-gun sound effects. After extensive testing, the machine-gun burst firing interval has been agreed upon and is programmed into each system.

Cue Light System
Since the actors perform the show to the music, a cue light system is used to indicate when certain actions, such as entrances, should occur. Large monitors backstage indicate what events are next; red lights next to the monitor come on for a warning and turn off to indicate that the event should take place. A similar system is implemented over the TD booth in the front of the house: A pair of green lights (one for backup in case of a burn out) mounted on top of the TD booth indicates to the performer playing Itchy that Scratchy is safely in place for certain effects, such as the grenade and the machine gun. The cue light control system accepts custom ASCII commands over Ethernet.

Audience Voting System
DGI contracts with a company that does audience voting systems to provide a phone-based system. The main controller for the audience voting system connects to the show control system over the show network to indicate the results of the voting using some custom ASCII messages; the show controller then handles the scheduling of the segments.

Park-wide Management
Park management systems will interface with the show controller through an Ethernet link, and this will allow the park's central control system to tell the show controller whether to run day or night versions of the show, and also do things such as turning on work lights at specific times for cleaning, and so on.

What system architecture works for this application?
While there are some primary-secondary relationships at times in the systems, it's generally designed as a distributed, peer-to-peer system.

What is the control information source?
Sometimes a time base will supply control information; other times the human operators will. The audience has a voting system to select and sequence segments, and, in critical safety situations, the actors onstage will actually provide additional control information themselves.

What type of user interface is required?
Skilled technicians are running the show and many elements of the show are unpredictable, so they will need very comprehensive user interfaces.

What kind of show control approach is needed?
This is a highly complex system that needs logic outside the connected systems, so we do need a dedicated show control system. To facilitate maintenance, training, and updates, Duff Gardens has standardized on a software-based system that runs on general-purpose computers. Since so much money is being spent on the attraction, DGI does extensive testing in advance to ensure that as few changes as possible are made after construction begins.

The show is broken up into segments, each of which is time-based, but stay flexible on any life-threatening or time-critical elements. Fifteen segments are available, but only nine (to represent Scratchy's nine lives) are to be selected by the audience for each show in order to keep the show interesting and generate repeat business. As the audience files into the venue, they select the first three segments they want to view, and in which order. After these segments, the "director" comes out and guides the audience through the selection of the next three.

Each segment is started by the technical director (TD) when everything is in place, safe, and operational. The show controller issues a go command to the music system, which then generates time code back on to the network, which in turn triggers the actors' cuing system, light cues, sound cues, and localization cues for that segment.

The system is a large network, with the systems connected as shown here:

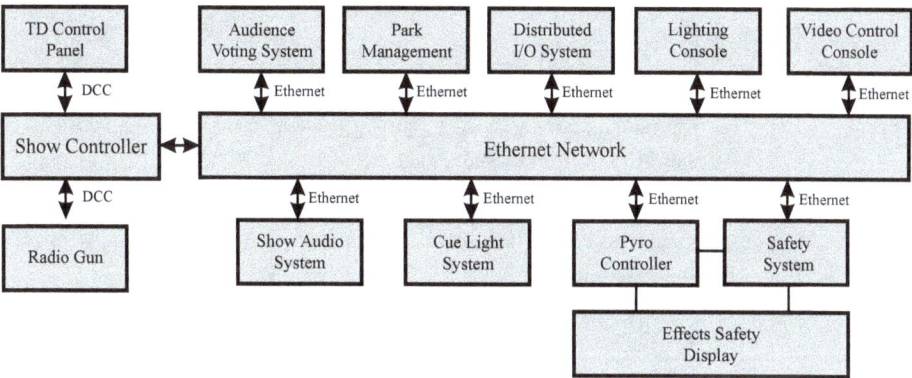

The show control programmers design this user interface:[5]

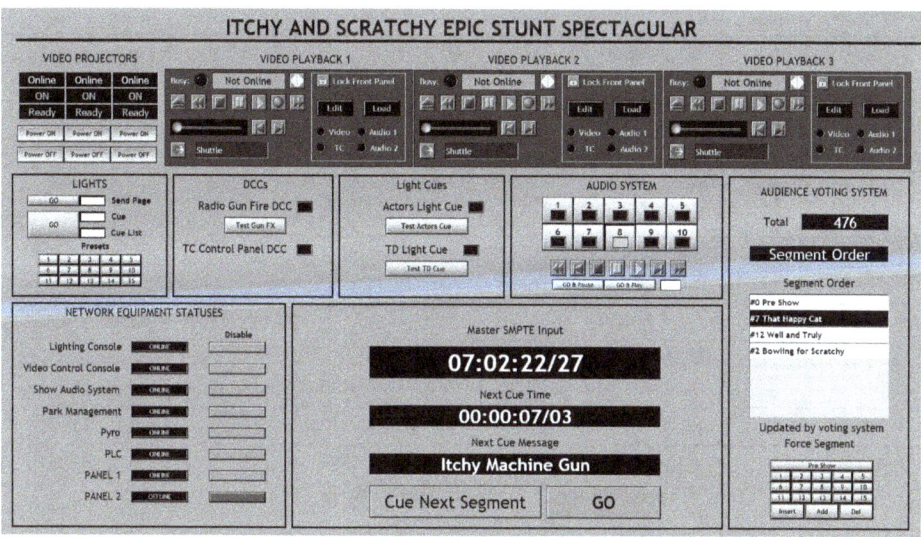

Show Control Script

DGI comes up with a detailed cue script for show planning. In addition to trigger sources and show control messages, the list indicates who executes a particular effect. "TD" denotes that the technical director executes the effect; "TC" indicates that the effect is executed automatically based on the time code; Snd means that the Sound Operator takes the cue; and Hs Mgr is the house manager in charge of the audience. This script applies to a nighttime show, including all lighting and image commands; for simplicity, however, cue-light cues, system acknowledgments, and other messages are not included here. For this example, the audience

---

5 User screen by Alan Anderson.

has selected segments 7, 12, and 2.

Time Code	Event	Actions	Ex By	Control Messages
Segment #0	**Preshow**			
Event	TD starts preshow	Park backgrnd music up Preshow light look Preshow images up	TD	Sound Go Lights Q000 Image Q000
Event	10 minutes before show	"Please take your seats, voting is about to begin"	Hs Mgr	Sound Q001
Event	2 minutes before show	"Please select the first three segments"	Hs Mgr	Sound Q002
Event	TD starts show	Itchy and Scratchy theme song	Snd	Sound Q005
00:00:10:13	2nd measure of music	Houselights to half	TC	Lights Q001
00:00:20:02	4th measure of music	Houselights out	TC	Lights Q002
00:00:24:28	Itchy in music	Itchy graphic up/ Preshow out	TC	Image Q001
00:00:32:07	Scratchy in music	Scratchy graphic up	TC	Image Q002
00:01:02:23	Start first segment		TC	
Segment #7	**That Happy Cat**			
07:00:00:00	Segment starts	Flowers music Go "That Happy Cat" image	TC TC	Music Q701 Image Q701
07:00:10:00	Scratchy enters	Lights for Scratchy up	TC	Lights Q701
07:00:18:00	Scratchy moves center	Sound localization cue Scratchy whistling	TC Snd	Localiz. Q702 Sound Q702
07:00:32:00	Itchy enters with flowers	Lights for Itchy up Itchy squeaking	TC Snd	Lights Q702 Sound Q703
07:02:30:00	Itchy enters with machine gun	TD authorizes machine gun	TD	
Event	Scratchy safely in place	Scratchy authorizes gun	Scr	
07:02:32:00	Scratchy safely in place	Lights for Scratchy in place	TC	Lights Q703
Event	Itchy squeezes trigger	Machine gun bullet hits Machine gun sounds Machine gun flashes Scratchy screaming	Itch Itch Itch Itch	Pyro Q701 Sound Q704 Lights Q704 Sound Q705

Time Code	Event	Actions	Ex By	Control Messages
Event	Scratchy breaks photocell beam	Flowers music fade out Lights for Scratchy off	TD TD	Sound Q706 Lights Q706
Event	Itchy releases trigger	Machine gun hits stop Machine gun sounds stop Machine gun flashes stop	Itch Itch Itch	Pyro Q702 Sound Q707 Lights Q705
Event	Itchy character laughs	Itchy laughing sound FX	Snd	Sound Q708
07:03:00:00	Itchy exits	Lights fade down	TC	Lights Q707
Segment #12	**Well and Truly**			
Event	Scratchy's head around corner	Well music go	TD	Sound Q1201
12:00:00:00	Scratchy staggers onstage	Lights for Scratchy up "Well and truly" image up	TC TC	Lights Q1201 Image Q1201
Event	Scratchy moaning	Moaning sound FX	Snd	Sound Q1202
Event	Takes drink from well	Slurping sound FX	Snd	Sound Q1203
12:00:30:00	Itchy enters	Lights for Itchy up	TC	Lights Q1202
Event	Itchy kicks Scratchy down well	TD authorizes explosion Falling sound FX	TD TD	Sound Q1204
12:00:45:00	Scratchy down well	Lights focus on well	TC	Lights Q1203
Event	Itchy laughs	Itchy Laughing sound FX	Snd	Sound Q1205
Event	Scratchy in safety position	Scratchy authorizes explosion	Scr	
Event	Itchy throws grenade	TD fires explosion Lights for explosion Sound for explosion Well music out	TD TD TD TD	Pyro Q1201 Lights Q1204 Sound Q1206 Sound Q1206
Event	Itchy laughs	Itchy laughing sound FX	Snd	Sound Q1207
12:02:05:00	Itchy walks offstage	Lights fade down	TC	Lights Q1205
Segment #2	**Bowling for Scratchy**			
Event	Scratchy's head peers out well	Bowling music go	TD	Sound Q201
02:00:00:00	Scratchy climbs out of well	Lights for Scratchy "Bowling for Scratchy" image Scratchy moaning	TC TC Snd	Lights Q201 Image Q201 Sound Q202
02:00:20:00	Scratchy lies down	Lights focus on Scratchy Purring sound FX Go	TC TC	Lights Q202 Sound Q203

Time Code	Event	Actions	Ex By	Control Messages
02:00:30:00	Itchy enters	Lights up on Itchy	TC	Lights Q203
Event	Itchy arranges bowling pins	Standby for explosion	TD	
02:01:45:00	Itchy arranges bowling pins	Lights focus on Itchy	TC	Lights Q204
Event	Scratchy moves to safety	Scratchy authorizes explosion	Scr	
Event	Itchy rolls bomb down alley	Bowling sound effects	Snd	Sound Q204
Event	Bomb ball hits pins	TD fires explosion Lights for explosion Sound for explosion Purring sound FX stop	TD TD TD TD	Pyro Q201 Lights Q205 Sound Q205 Sound Q206
02:03:00:00	Itchy walks offstage	Itchy laughing sound FX	TC	Sound Q207

Before each show, the pyrotechnicians load all the pyro equipment and walk through the attraction with the stunt performers to do a safety check. When the show is ready for the audience, the TD executes a cue on the controller that starts pre-show mode. In this mode, special park background music is piped through the system, along with any park-wide paging or announcements. If the night show is being run, a pre-show light look is brought up, and the Itchy and Scratchy image is projected on the water tower. The house manager keeps in touch with the TD as the audience files in, and the Sound Operator triggers one announcement 10 minutes before show time: "Ladies and gentlemen, please take your seats. The Itchy and Scratchy Stunt Show will start in 10 minutes. You can start voting now for your first three stunts." Two minutes before show time, the house manager checks again and the Sound Op triggers another recorded announcement. At this point, the TD gets verbal confirmation over the headset, closes the voting (which loads up the first three segments), confirms that the actors are ready, and calls places. When the house manager indicates that the audience is seated, the Sound Op starts the introduction sequence—the ever-popular Itchy and Scratchy theme.

As the houselights dim, strains of the theme song can be heard over the sound system: "They fight, they fight, they fight, they fight, they fight! Fight-fight-fight, fight-fight-fight! The Itchy and Scratchy Epic Stunt Spectacular!" The show controller generates time code, which is distributed over the network, which in turn triggers the video system, and the recorded announcer introduces the popular animated characters. At time code 00:01:02:23, on the final downbeat, the first audience selected segment (#7 in this case) is automatically triggered.

The first music cue begins, the image system brings up the title of the first segment—"That Happy Cat". At time 07:00:05:00, Scratchy's cue light comes on; at 07:00:10:00 the light goes off and Scratchy enters. The Sound Op triggers a whistling cue as the performer turns their head toward the audience. Itchy is cued to enter and comes onstage with their bouquet of flowers. The Sound Op executes a dialogue cue, and Itchy walks offstage.

When Itchy enters with the machine gun and the TD sees that Itchy is in the right position and that everything is safe, they authorize the machine gun sequence. Scratchy moves to the correct position to go flying through a breakaway wall when hit by the bullets; to signal that they are ready, and stands on two switches mounted in the deck, one for each foot. These switches are also wired into the pyro safety system, and when both switches are closed, the system closes a contact to the pyro controller, which is indicated on the TD's control panel and received by the show controller. When the show controller receives this message, the green lights on top of the booth come on, indicating to Itchy that everything is ready. As soon as Itchy pulls the trigger, a radio signal is sent, which (through the show control system) triggers the machine-gun sound effects, special lighting, a scream sound for Scratchy, and the bullet hits. Scratchy crashes through the wall, and the Sound Op executes a cue that fades the flowers music out (although time code is still generated). The time code turns Scratchy's light look off. When Itchy releases the trigger, all the machine-gun effects stop. Itchy moves their head in a way to signal the Sound Op to play a sadistic laugh cue, and then walks offstage. A light cue is taken, leaving only a spotlight on the hole in the wall where Scratchy crashed through.

When the audience sees Scratchy's bullet-ridden head peer around the corner, they give a big cheer, and the Sound Op starts the second segment (#12 in this case), starting the "well" music; time code then starts the other elements. Scratchy moves toward the well, several light and localization cues are triggered automatically, and the Sound Op triggers a moan sound. As Scratchy leans into the well, the Sound Op triggers several manual slurping sounds; then Itchy's cue light goes out; Itchy enters, and kicks Scratchy down the well. The TD executes a cue that puts the explosion effect into standby and the Sound Op triggers a falling and splash sound effect.

The performer playing Scratchy, of course, only falls about six feet and climbs into a special bunker to be protected from the explosion. Itchy dances around the top of the well to cover the time Scratchy needs to get into the safety position. When in place, Scratchy presses and holds an authorize switch in a bunker underneath the well, which signals the pyro controller, the TD, and the show controller, which then turns on the green lights on top of the booth. Itchy sees the green ready light and tosses the hand grenade down the well; the TD waits a moment and then executes a cue triggering the water-cannon effect, a light cue, an explosion sound effect, and a cue that fades out the "well" music. The Sound Operator triggers a laughing cue, and Itchy walks offstage. The lights fade down, and finally Scratchy's charred head is seen over the top of the well; the TD starts the next sequence.

The Sound Op triggers appropriate moaning cues, and other cues are taken automatically from the time code as Scratchy staggers over to the bowling alley. The Sound Op executes a purring cue as Scratchy lays down; Itchy's cue light goes off, and Itchy enters. Itchy's act of arranging the large bowling pins is actually a diversion as Scratchy is moving to an explosion-proof bunker behind the alley. The TD executes a cue telling the pyro system to stand by, Scratchy presses and holds a safety switch, and the green cue light comes on. Itchy sees the light and rolls the ball down the alley. As the ball hits the pins, the TD triggers a command firing the huge explosion cue, the sound effect, light cue, and a sound cue that kills the purring sound. As Itchy walks offstage, the Sound Operator triggers another laughing cue.

We've covered only the first three of nine segments of the show, but the attraction is a rousing success, and people line up all day to get in! Mr. Burns says the show is "*Ex-cellent!*"

CONCLUSION

Many aspects of the 1980s world of entertainment technology in which I started my career are gone. Back then, most shows used simple approaches to link a few devices; more complex, powerful systems had to be custom engineered, and this meant that powerful technologies were available in limited ways only to theme parks and similar shows and attractions with large budgets and lots of resources. I'm not a nostalgic person, and so I say good riddance to the old days. The affordable availability of phenomenal computer power, coupled with the common data highway of Ethernet, has now leveled the playing field, distributing these amazing technologies to ever-smaller kinds of production. Low-budget shows now have creation power and capabilities in a laptop that would have only been available to theme park or concert designers in the 1990s.

I have also argued[1] that show technology has stabilized and matured, and this is a good thing, since it allows us to focus more on telling stories instead of inventing tools—we no longer have to invent the paint brush to make a painting. And show control, leveraging cheap computing power and the connectivity of networks, is now available to anyone with an interest, and is a key opportunity for future practical innovation. I think today we are no longer significantly restricted by technology, but instead are only limited by the laws of nature, budget and our creative imaginations. This is something I couldn't have written in my first book in 1994.

CONTACT INFO AND BLOG

If you have any comments, questions, or corrections, I'd like to hear them. Please check my website at: http://www.controlgeek.net You can contact me from there, and I have a blog, book errata, and lots of other information.

Thanks!

John Huntington
NYC, May 2023

---

[1] See the articles I wrote here https://controlgeek.net/blog/2020/8/20/development-and-evolution-of-show-technology-articles-and-timeline

INDEX

A

absolute 10, 11, 12, 13, 32, 43
AHJ. See Authority Having Jurisdiction
algorithms 3
Amazon.com xiii
American Standard Code for Information Interchange 24, 25, 26, 78, 79, 80, 85, 86, 97
Anderson, Alan 82, 89, 99
animatronic 17, 76, 80
Art-Net 40
ASCII. See American Standard Code for Information Interchange
audio 16
audio distribution amplifier 35
Authority Having Jurisdiction 49, 62

B

bandwidth 3
binary xiv, 3, 24, 41
bit rate 3
bits per second 3
BLE. See Bluetooth Low Energy
blog xiv, 107
Bluetooth Low Energy 22
Bollinger, Aaron xv
bowling 77
BPS, see bits per second
Burns, Mr. 95

C

carriage return 24
chase 32, 38
circus 1
click track 39, 87, 88
client-server 15
closed loop 10
command format 29
concert 1
conductor-follower 13
contact closure 41
contactor 44
containers 26
controller-responder 13

Control Systems for Live Entertainment xiii
COVID xiv
cruise ship 1
cue 4
cue light 97
cue list 30
cue path 30

D

DA. See Distribution Amplifier
dark operate 43
DCC. See dry contact closure
dead man's switch 52
device ID 30
DMX512 xiii, 23, 35, 39
Drama Book Shop xiii
driver's safety device 52
drop frame 37
dry contact closure 41, 74, 76
Duff Gardens 98

E

elegant 56
emergency stop 50
enabling 51
encoder 44
entertainment control systems xiii, xv, 1, 3, 5, 21
entertainment technology 1
E-Stop. See emergency stop
Ethernet 2, 24, 35
explosion 6, 7, 8, 9, 10, 94, 97, 101, 102, 103, 104

F

fail-safe design 50
fire 18
fire alarm 18
firewall 57
fireworks 7, 33, 37
FMEA. See Failure Mode and Effects Analysis
Focal Press xiii
fog machine 1
freewheeling 38

G

General Purpose Interface 41

gigabit 3
Google xiii
GPI. See General Purpose Interface
graceful abort 52

H

hackers xiv
Hackers On Planet Earth xiv
hazard analysis 52
host 2
hot backup 53
HVAC 18

I

immersive 1
Internet of Things 46
Internet Protocol 2, 3, 25, 31, 68, 81
interrupt-driven 61
Introduction to Show Control xiii, xiv, 2, 3, 25, 31, 68, 81
Introduction to Show Networking xiv, 2, 22, 24, 25, 41, 54, 57
IoT. See Internet of Things
IP. See Internet Protocol
IP Address 2
isolation 45
isolation, galvanic 45
isolation, optical 45

J

jam-sync 36

K

keyboard 28

L

laser 17
Lawrence, Michael xv
lighting xv, 16, 39, 50
Lighting Dimensions xiii
light operate 43
limit switch 43
linear show 5
Linear Time Code 35, 36, 37, 67. See also time code
longitudinal 35
LTC. See Linear Time Code

M

main-secondary 13
Mbit/s 3
megabit 3
methods 26
metronome 39
MIDI. See Musical Instrument Digital Interface
MIDI Show Control 16, 17, 29, 30, 31, 32, 70, 71, 74, 76, 85, 87
MIDI Time Code 34, 36, 37
motion-control 80
motion detectors 43
moving light 1
MSC. See MIDI Show Control
MTC. See MIDI Time Code
museum 1
Musical Instrument Digital Interface xiii, 16, 17, 21, 26, 27, 28, 29, 30, 31, 32, 34, 36, 37, 47, 66, 70, 71, 72, 73, 74, 76, 80, 81, 85, 86, 87, 88
musical time 9

N

network 1, 22
Network Time Protocol 46
node 2
non-drop 37
nonlinear show 5
NTP. See Network Time Protocol
NTSC 34
NYC ii, 107

O

Object Transform Protocol 46
open loop 10
Open Sound Control 16, 17, 21, 25, 26, 27, 28, 29, 30, 31, 32, 34, 36, 37, 47, 66, 70, 71, 72, 73, 74, 76, 79, 80, 81, 85, 86, 87, 88
Operator Presence Control 52
OSC. See Open Sound Control
OTP. See Object Transform Protocol

P

peer-to-peer 14, 60, 72, 80, 87, 92, 98
point-to-point 21, 22, 39
polling 14
Popsicle kitteh 90
PosiStageNet 46
Precision Time Protocol 47

pre-roll 37, 67, 68, 86
preset/command 22
primary-secondary 13, 14, 60, 66, 98
producer-consumer 15
programming 3
protocol 22
PTP. See Precision Time Protocol
publisher-subscriber 15
pyro 1, 5, 18

R

RDM. See Remote Device Management
RDMnet 40
Real Time Tracking Protocol 46
redundancy 53
Reese, Shelbye xv
relative 10, 11, 12, 37, 56
relay 44, 50, 74
Remote Device Management 40
repetitive/continuous 22
retro-reflective 43
RS-232 21
RS-422 21
RS-485 21
RTTrP. See Real Time Tracking Protocol

S

sACN. See streaming ACN
safe torque off 50
safety 49, 58
safety cable 50
safety relay 50
sensor 42, 43, 44, 91
sensors 42
serial 21, 22
servers 1, 2, 15, 17, 25, 33, 39, 78, 86
show business 1
show control xv, 1
Show Networks and Control Systems xiv
Simple Network Management Protocol 46
single failure-proof design 50
slots 39
smoke 18
SMPTE. See SMPTE Time Code
SMPTE Time Code xiii, 35, 79, 88. See also time code
SNMP. See Simple Network Management Protocol

stage machinery 17
stage manager 9
storytelling 1
streaming 2
streaming ACN 16, 40, 86
switch 41
switches 42
synchronization 3, 7
synchronize 1
systems thinking 4

T

tape 35, 38
TCP. See Transmission Control Protocol
theater 1, 70, 71
Theatre Crafts xiii
theme park 1
time-based 7, 8, 9, 38, 59, 66, 70, 87, 95, 98
time code xiii, 7, 13, 14, 16, 17, 18, 32, 33, 34, 35, 36, 37, 38, 39, 46, 60, 66, 67, 68, 78, 79, 80, 81, 85, 86, 87, 88, 98, 99, 100, 101, 102, 103, 104
Times Square xiii
Tomaino, Daniela 92
tornado 65
transistor 41, 45
Transmission Control Protocol 25

U

UDP. See User Datagram Protocol
Unicode 25
Universal Serial Bus 22
universe 39
USB. See Universal Serial Bus
User Datagram Protocol 25
UTF-8 24

V

value engineering 62
velocity 28
Vertical Interval Time Code 35
video 17
VITC. See Vertical Interval Time Code

W

water 18
website xv, 107
wild 7, 38

wrestling 1

Z

Zigbee 22
Zircon Designs ii
Zircon Designs Press ii

www.ingramcontent.com/pod-product-compliance
Lightning Source LLC
Chambersburg PA
CBHW050254120526
44590CB00016B/2341